# Stability of Spherically Symmetric Wave Maps

MEMOIRS
of the
American Mathematical Society

Number 853

# Stability of Spherically Symmetric Wave Maps

Joachim Krieger

May 2006 • Volume 181 • Number 853 (second of 5 numbers) • ISSN 0065-9266

**American Mathematical Society**
Providence, Rhode Island

2000 *Mathematics Subject Classification.*
Primary 35L05, 35L70.

**Library of Congress Cataloging-in-Publication Data**

Krieger, Joachim, 1976–
Stability of spherically symmetric wave maps / Joachim Krieger.
p. cm. — (Memoirs of the American Mathematical Society, ISSN 0065-9266 ; no. 853)
"Volume 181, number 853 (second of 5 numbers)."
Includes bibliographical references.
ISBN 0-8218-3877-6 (alk. paper)
1. Wave equation. 2. Differential equations, Parabolic. I. Title. II. Series.

QA3 .A57 no. 853
[QA174.26.W28]
510 s—dc22
[530.12′4] 2006040673

## Memoirs of the American Mathematical Society

This journal is devoted entirely to research in pure and applied mathematics.

**Subscription information.** The 2006 subscription begins with volume 179 and consists of six mailings, each containing one or more numbers. Subscription prices for 2006 are US$624 list, US$499 institutional member. A late charge of 10% of the subscription price will be imposed on orders received from nonmembers after January 1 of the subscription year. Subscribers outside the United States and India must pay a postage surcharge of US$31; subscribers in India must pay a postage surcharge of US$43. Expedited delivery to destinations in North America US$35; elsewhere US$130. Each number may be ordered separately; *please specify number* when ordering an individual number. For prices and titles of recently released numbers, see the New Publications sections of the *Notices of the American Mathematical Society.*

**Back number information.** For back issues see the *AMS Catalog of Publications.*

Subscriptions and orders should be addressed to the American Mathematical Society, P. O. Box 845904, Boston, MA 02284-5904, USA. *All orders must be accompanied by payment.* Other correspondence should be addressed to 201 Charles Street, Providence, RI 02904-2294, USA.



*Memoirs of the American Mathematical Society* is published bimonthly (each volume consisting usually of more than one number) by the American Mathematical Society at 201 Charles Street, Providence, RI 02904-2294, USA. Periodicals postage paid at Providence, RI. Postmaster: Send address changes to Memoirs, American Mathematical Society, 201 Charles Street, Providence, RI 02904-2294, USA.



This publication is indexed in *Science Citation Index*®, *SciSearch*®, *Research Alert*®, *CompuMath Citation Index*®, *Current Contents*®/*Physical, Chemical & Earth Sciences.*
Printed in the United States of America.

♾ The paper used in this book is acid-free and falls within the guidelines established to ensure permanence and durability.
Visit the AMS home page at http://www.ams.org/

10 9 8 7 6 5 4 3 2 1 11 10 09 08 07 06

# Contents

# Abstract

We study Wave Maps from $\mathbf{R}^{2+1}$ to the hyperbolic plane $\mathbf{H}^2$ with smooth compactly supported initial data which are close to smooth spherically symmetric initial data with respect to some $H^{1+\mu}$, $\mu > 0$. We show that such Wave Maps don't develop singularities in finite time and stay close to the Wave Map extending the spherically symmetric data(whose existence is ensured by a theorem of Christodoulou-Tahvildar-Zadeh) with respect to all $H^{1+\delta}, \delta < \mu_0$ for suitable $\mu_0(\mu) > 0$. We obtain a similar result for Wave Maps whose initial data are close to geodesic ones. This strengthens a theorem of Sideris for this context.

---

Received by the editor October 30, 2004.

2000 *Mathematics Subject Classification.* Primary 35L05, 35L70.

Author partially supported by NSF grant DMS-0401177.

# Preface

This paper grew out of an attempt to understand the relation between the techniques developed by Christodoulou and Tahvildar-Zadeh in their deep study of large spherically symmetric Wave Maps (in $2+1$-d) and the recently developed techniques from harmonic analysis that led to general long-time existence and regularity preservation results by D. Tataru, T. Tao, as well as the author in the context of $2+1$-d Wave Maps, provided the data are small with respect to energy but not necessarily spherically symmetric. While the analysis of Christodoulou-Tahvildar-Zadeh takes place mostly in physical space, deducing pointwise estimates on the Wave Maps and their derivative components, the harmonic analytic treatment so far has mostly taken place purely in frequency space. In this paper, this discrepancy is overcome by using some sort of wave packets, thus using localization both in frequency and physical space, to deduce the stability of spherically symmetric Wave Maps. While a frequency-based analysis works in the regime of extremely low or high frequencies, the pointwise decay estimates become useful in the regimes of moderate frequencies. Thus the two approaches are seen to beautifully complement each other. It is to be hoped that these techniques can be pushed further to approach the problem of regularity preservation and stability for general large-energy Wave Maps to hyperbolic targets.

Joachim Krieger

CHAPTER 1

# Introduction, Controlling Spherically Symmetric Wave Maps

## 1.1. Introduction

Let $\{(\mathbf{x},\, \mathbf{y})|\mathbf{y} > 0\}$ be the hyperbolic plane, equipped with metric $dg = \frac{d\mathbf{x}^2+d\mathbf{y}^2}{\mathbf{y}^2}$, and let $\mathbf{R}^{n+1}$, $n \geq 1$ denote the standard Minkowski space equipped with the metric $dh = -dx_0^2 + \sum_{i=1}^n d_{x_i}^2$. We shall also use the identifications $x_0 = t$ (time), $\partial_{x_\nu} = \partial_\nu$, $\nu = 0, 1, \dots, n$. A Wave Map from Minkowski space to $\mathbf{H}^2$ is a map $u : \mathbf{R}^{n+1} \to \mathbf{H}^2$ which is critical with respect to the functional

$$u \to \int_{\mathbf{R}^{n+1}} < \partial_\alpha u, \partial^\alpha u >_g dx_0 d_{x_1} \dots d_{x_n},$$

where $\partial_\alpha u = u_*(\partial_\alpha)$, $\partial^\alpha = h^{\alpha\beta}\partial_\beta$ and Einstein's summation convention is in force. The Euler Lagrange equations of this problem read as follows:

$$\Box \ln \mathbf{y} = -\frac{\partial_\nu \mathbf{x}}{\mathbf{y}} \frac{\partial^\nu \mathbf{x}}{\mathbf{y}} \tag{1.1}$$

$$\Box(\frac{\mathbf{x}}{\mathbf{y}}) = \frac{\mathbf{x}}{\mathbf{y}} \frac{\partial_\nu \mathbf{y} \partial^\nu \mathbf{y} + \partial_\nu \mathbf{x} \partial^\nu \mathbf{x}}{\mathbf{y}^2} \tag{1.2}$$

If $n = 2$, the fundamental Conjecture associated with this problem is the following, which flows from the intuition that the negative curvature of the target should prevent a focusing of energy in small spatial regions:

**Conjecture**(e. g. Klainerman [**10**]) Let $n = 2$. Given smooth initial data $(\mathbf{x}, \mathbf{y})$, $(\partial_t \mathbf{x}, \partial_t \mathbf{y}) : \mathbf{R}^2 \times \{0\} \to \mathbf{H}^2 \times T\mathbf{H}^2$, there exists a global-in-time smooth Wave Map extending them.

This is expected to be generalizable to arbitrary targets of negative curvature and satisfying some geometric niceness conditions. We stick in this paper to the $\mathbf{H}^2$ model on account of its simplicity.
The difficulty in establishing the above conjecture stems from the fact that the problem is energy critical, i. e. the natural scale invariant Sobolev space is exactly the energy space $\dot{H}^1$ (the energy $\sum_{\alpha=0}^n ||\frac{\partial_\alpha \mathbf{x}}{\mathbf{y}}||_{L_x^2}^2 + ||\frac{\partial_\alpha \mathbf{y}}{\mathbf{y}}||_{L_x^2}^2$ is preserved under the Wave Map flow). Establishing global regularity for such problems consists customarily of showing that smooth small data imply global regularity, as well as non-concentration of energy in physical space. The latter needs to depend subtly on the geometry of the target, since a priori analytic reasons cannot rule out a rapid shift of the energy from low to high frequency modes, resulting in sudden focusing. In the case $n = 3$, one expects breakdown of solutions for large data for analytic reasons (the scale invariant Sobolev space $\dot{H}^{\frac{3}{2}}$ which in some sense controls the

local well-posedness behavior is not controlled by the energy). We can formulate

**Conjecture:** *Let* $n = 3$. *There exist (large) smooth initial data* $(\mathbf{x}, \mathbf{y})$, $(\partial_t \mathbf{x}, \partial_t \mathbf{y}) : \mathbf{R}^3 \times \{0\} \to \mathbf{H}^2 \times T\mathbf{H}^2$, *which lead to breakdown in finite time.*

Breakdown solutions are known in 3+1 dimensions, but only for special targets [**3**] not including the hyperbolic plane.
The best result known at this point pertaining to the first Conjecture is the following theorem of the author [**22**], [**23**]:

THEOREM 1.1. *Let* $n = 2, 3, \ldots$. *Then there exists* $\epsilon > 0$ *such that for smooth initial data* $(\mathbf{x}, \mathbf{y})$, $(\partial_t \mathbf{x}, \partial_t \mathbf{y}) : \mathbf{R}^n \times \{0\} \to \mathbf{H}^2 \times T\mathbf{H}^2$ *satisfying*

$$\int_{\mathbf{R}^2} \sum_{\nu=0}^{n} ||\frac{\partial_\nu \mathbf{x}}{\mathbf{y}}||_{\dot{H}^{\frac{n}{2}}} + ||\frac{\partial_\nu \mathbf{y}}{\mathbf{y}}||_{\dot{H}^{\frac{n}{2}}} < \epsilon,$$

*there exists a smooth global-in-time Wave Map extending them.*

This is similar to earlier results of Tao [**37**] when the target is a sphere, as well as of Tataru [**41**] when the target is uniformly isometrically embeddable into a Euclidean space. Similar results in dimensions $n \geq 4$ for quite general targets were achieved by Klainerman-Rodninanski[**15**], Shatah-Struwe[**27**] as well as Nahmod-Stefanov-Uhlenbeck[**24**] after Tao's initial breakthrough [**36**], [**37**].
Thus the preceding theorem does not yet exhibit behavior reflecting the geometric nature[1] of $\mathbf{H}^2$.

What we set out to do in this paper is to try to exploit features which appear to hinge on geometric properties of this target and set it apart from positively curved targets such as the sphere $S^2$. We need the following definition:

DEFINITION 1.2. : We call a Wave Map $u : \mathbf{R}^{n+1} \to M$ 'spherically symmetric' provided $u(t, \rho x) = u(t, x)$ $\forall x \in \mathbf{R}^2$ and $\rho : S^1 \to SO(2)$ the standard representation of $S^1$ on $\mathbf{R}^2$.

We shall use the deep results of Christodoulou-Tahvildar-Zadeh[**5**] on the asymptotic behavior of spherically symmetric Wave Maps, valid for certain targets which amongst other things have no conjugate points, to conclude the following:

THEOREM 1.3. *(Stability) Let* $n = 2$ *and let* $u : \mathbf{R}^{2+1} \to \mathbf{H}$ *be a smooth spherically symmetric Wave Map with compactly supported initial data* $u[0] = (u(0), \partial_t u(0)) : \mathbf{R}^2 \times \{0\} \to \mathbf{H}^2 \times T\mathbf{H}^2$. *Then for any* $\sigma > 0$ *there exists a number* $\epsilon > 0$ *such that for all initial data* $\tilde{u}[0]$ *which are* $\epsilon$*-close to* $u[0]$ *in* $H^{1+\sigma}$, *there exists a smooth global in time Wave Map* $\tilde{u}$ *extending* $\tilde{u}[0]$. *Moreover,* $\tilde{u}$ *will stay close to* $u$ *in the energy topology*[2] *(in a suitable sense)globally in time.*

This is a type of 'large data result', although of course there is still a smallness assumption present. As far as the case of target $S^2$ is concerned, a surprising result of M. Struwe [**34**] asserts that smooth radial data lead to global Wave Maps. This

[1]Paradoxically, the proof of this result involved extra complications over the case of target a sphere, on account of the fact that one needs to work with the derivative of the Wave Map, losing one degree of smoothness.

[2]Indeed, even in a certain range of subcritical spaces $H^{1+\lambda}$.

suggests the important question of whether these solutions are unstable:

**Question***(Instability?)*. *Let $u[0] : \mathbf{R}^2 \times \{0\} \longrightarrow S^2 \times TS^2$ be large generic spherically symmetric initial data. Is it true that for any $\sigma > 0$, $\epsilon > 0$, there exist smooth initial data $\tilde{u}[0] : \mathbf{R}^2 \times \{0\} \longrightarrow S^2 \times TS^2$ with the property that $||u[0] - \tilde{u}[0]||_{H^{1+\sigma}} < \epsilon$ while[3] the smooth Wave Map $\tilde{u}$ extending $\tilde{u}[0]$ locally in time breaks down after finite time? More precisely for any $\delta > 0$*

$$\exists T < \infty \rightarrow \forall \delta > 0 \lim_{t \rightarrow T_-} ||\tilde{u}[t]||_{H^{1+\delta} \times H^\delta} = \infty$$

Unfortunately, our techniques appear to have no bearing on this question. For example, we don't know what the asymptotic behavior of Struwe's solutions is.

The key ingredient to prove Theorem 1.3 is the boundedness of a range of subcritical Sobolev norms for large spherically symmetric Wave Maps:

THEOREM 1.4. *: There exists $\delta_0 > 0$ such that $\forall 0 \leq \delta < \delta_0$ and spherically symmetric smooth Wave Maps $u : \mathbf{R}^{2+1} \longrightarrow \mathbf{H}^2$ we have*

$$\sup_t ||u[t]||_{H^{1+\delta} \times H^\delta} < \infty$$

The proof of this will follow from the asymptotic results of [**5**], which in turn rely on a careful analysis of conservation laws associated with (1.1), (1.2).
We shall then rely on the setup of [**23**], using the intrinsic derivative formulation (by differentiating (1.1), (1.2)) and passing to the Coulomb Gauge. The new difficulties by comparison with [**23**] concern nonlinear terms which are linear in the difference $\nabla[\tilde{u} - u]$. Working in the Coulomb Gauge, this corresponds to perturbing the flat d'Alembertian $\Box$ with a potential term $V$ which is in some sense quadratic in the derivatives of the spherically symmetric Wave Map. We shall show that the good decay behavior of the spherically symmetric Wave Map allows us to treat these terms as source terms, instead of having to modify the linear operator. However, the fact that we cannot just work with mixed Lebesgue type norms but complicated null-frame spaces will force simultaneous localizations in physical and frequency space on us, which make the argument quite intricate. These types of estimates might be useful when working on the general large data problem. Our analysis shall have as simple corollary a generalization of a result of Sideris[**29**] to $n = 2$: we define a geodesic Wave Map $u(t, x)$ to be of the form $u = \gamma(v)$ where $\gamma(.) : \mathbf{R} \rightarrow \mathbf{H}^2$ is a geodesic and $\Box v = 0$. Then we have the following:

THEOREM 1.5. *Let $u(t, x) : \mathbf{R}^{2+1} \longrightarrow \mathbf{H}^2$ be a smooth geodesic Wave Map. Then there exists $\epsilon > 0$ such that for all initial data $\tilde{u}[0]$ $\epsilon$-close to $u[0]$ in $H^{1+\sigma}$, there exists a global Wave Map extending $\tilde{u}[0]$. Also , $\tilde{u}$ will stay close to $u$ in the energy topology in a suitable sense.*

We also point out that due to a result of Shatah-Tahvildar-Zadeh [**26**] on the asymptotic behavior of equivariant Wave Maps, one expects a similar result for perturbations of large equivariant Wave Maps to hyperbolic targets.

[3]To define this norm, use $S^2 \hookrightarrow \mathbf{R}^3$ and use standard coordinates

### 1.2. A priori estimates for spherically symmetric Wave Maps.

For a Wave Map $u = (\mathbf{x}, \mathbf{y}) : \mathbf{R}^{2+1} \to \mathbf{H}^2$, we define the norm $||u(t)||_{H^s}$ as

$$||(\mathbf{x}, \mathbf{y})||_{H^s} := \sum_{\nu=0}^{2} ||\frac{\partial_\nu \mathbf{x}}{\mathbf{y}}(t)||_{H^{s-1}} + ||\frac{\partial_\nu \mathbf{y}}{\mathbf{y}}||_{H^{s-1}}$$

We also introduce the following notation: $r = \sqrt{x_1^2 + x_2^2}$. Now let $u(t,x) = (\mathbf{x}, \mathbf{y})$ be a spherically symmetric Wave Map with compactly supported smooth initial data. Then we have

LEMMA 1.6. *The image of the Wave Map belongs to a bounded subset of* $\mathbf{H}^2$*. More precisely, we have*

$$||\ln \mathbf{y}||_{L_t^\infty L_x^\infty} < \infty, \; ||\frac{\mathbf{x}}{\mathbf{y}}||_{L_t^\infty L_x^\infty} < \infty$$

*The bounds depend (at most) on the size of the support as well as some norm* $||u[0]||_{H^{1+\delta}}$, $\delta > 0$.

PROOF. : We shall rely on the following Proposition in [**5**]:

PROPOSITION 1.1. *(Chr.-Tah) Under the previous assumptions, the following inequalities hold:*

$$\text{'Good derivative'}: \; |\frac{\partial_v \mathbf{x}}{\mathbf{y}}| + |\frac{\partial_v \mathbf{y}}{\mathbf{y}}| \leq C(t+r)^{-\frac{3}{2}}, \; \partial_v = \partial_t + \partial_r$$

$$\text{'Bad derivative'}: \; |\frac{\partial_u \mathbf{x}}{\mathbf{y}}| + \frac{\partial_u \mathbf{y}}{\mathbf{y}}| \leq C(|t-r|+1)^{-1}(t+r)^{-\frac{1}{2}}, \; \partial_u = \partial_t - \partial_r$$

By local well-posedness of (1.1), (1.2) in $H^s$, $s > 1$, there exists a time interval $[-T, T]$ with $T = T(||u[0]||_{H^s})$ on which we uniformly control $||\ln \mathbf{y}(t)||_{H^s}$, $t \in [-T, T]$. Given an arbitrary point $(t, x) \in \mathbf{R}^{2+1}$ at distance $< \frac{T}{\sqrt{2}}$ from the forward light cone (say), connect it to a point in the strip $[-T, T] \times \mathbf{R}^2$ by means of a null-geodesic $\gamma$ given by $t - r = u = \text{const}$. We have

$$|\frac{\partial_v \mathbf{y}}{\mathbf{y}}| \leq C(t+r)^{-\frac{3}{2}}$$

on $\gamma$, whence

$$\int_{\gamma \cap \{t \geq T\}} |\frac{\partial_v \mathbf{y}}{\mathbf{y}}| dt < \infty$$

Combining this with the embedding $H^s \subset L_x^\infty$ yields the claim for $\ln \mathbf{y}$ for such points $(t, x)$. Given a point $(t, x)$ at distance $\geq \frac{T}{\sqrt{2}}$ from the forward light cone, connect it via a geodesic $\gamma$: $t + r = \text{const}$ to a point in the strip of thickness $\frac{T}{\sqrt{2}}$ around the light cone. Using

$$|\frac{\partial_u \mathbf{y}}{\mathbf{y}}| \leq C u^{-1} v^{-\frac{1}{2}}$$

yields

$$\int_\gamma |\frac{\partial_u \mathbf{y}}{\mathbf{y}}| dt < \infty$$

Thus the claim follows in general for $\ln \mathbf{y}$. With this, one proceeds similarly for $\mathbf{x}$. $\square$

The following is the main result of this section:

PROPOSITION 1.2. *Let the assumptions be as in the preceding lemma. There exists $\epsilon_0 > 0$ such that $\forall 0 < \epsilon < \epsilon_0$, there exists a global bound*

$$\sum_{\nu=0}^{2} ||\partial_\nu \ln \mathbf{y}||_{L_t^\infty H^\epsilon} < C_\epsilon, \sum_{\nu=0}^{2} ||\frac{\partial_\nu \mathbf{x}}{\mathbf{y}}||_{L_t^\infty H^\epsilon} < C_\epsilon,$$

*where $C_\epsilon$ depends on the size of the support as well as the $L^2$-mass of finitely many derivatives of $u[0]$*[4]

PROOF. : We shall need the following lemma, which is also due to Christodoulou-Tahvildar-Zadeh:

LEMMA 1.7. *(Chr.-Tah) Introduce the vector fields $S = t\partial_t + r\partial_r$, $\Omega = r\partial_t + t\partial_r$. For a smooth spherically symmetric Wave Map $u(t, x) : \mathbf{R}^{2+1} \to \mathbf{H}^2$ and $\epsilon > 0$, we have*

$$r^{\frac{1}{2}+\epsilon} v^{1-\epsilon}[|\partial_v S\mathbf{x}| + |\partial_v S\mathbf{y}|] < \infty$$

$$r^{\frac{1}{2}+\epsilon} v^{1-\epsilon}[|\partial_v \Omega\mathbf{x} + |\partial_v \Omega\mathbf{y}|] < \infty$$

*As a consequence, we conclude that*

$$|\nabla_{x,t}\partial_v \mathbf{x}| + |\nabla_{x,t}\partial_v \mathbf{y}| \leq C v^{-(1-\epsilon)} u^{-1} r^{-\frac{1}{2}-\epsilon}$$

We shall in fact prove that the quantity

$$A(t) := \sum_{k\in\mathbf{Z}} 2^{\delta|k|}[||P_k(\ln \mathbf{y}(t,.))||_{\dot{H}_x^1} + ||P_k(\frac{\nabla_x \mathbf{x}}{\mathbf{y}}(t,.))||_{L_x^2}$$
$$+ ||P_k\partial_t(\ln \mathbf{y})(t,.)||_{L_x^2} + ||P_k(\frac{\partial_t \mathbf{x}}{\mathbf{y}})(t,.)||_{L_x^2}]$$

is bounded globally in time, provided $\delta > 0$ is chosen sufficiently small. We have introduced here the Littlewood-Paley multipliers $P_k$, $k \in \mathbf{Z}$, which localize the (spatial) Fourier support to dyadic size $|\xi| \sim 2^k$. More precisely, let $\chi_0(.) \in C_0^\infty(\mathbf{R}_{>0})$ have support contained in $(\frac{1}{2}, 2)$ and satisfy

$$\sum_{j\in\mathbf{Z}} \chi_0(\frac{x}{2^j}) = 1 \, \forall x \in \mathbf{R}_{>0}.$$

Then we define (see also [**31**]) $P_k f$ for any $f \in L^1(\mathbf{R})$ via

$$\widehat{P_k f}(\xi) = \chi_0(\frac{|\xi|}{2^k})\hat{f}(\xi)$$

We need to show that $A(T) \leq C1 + \int_0^T A(t)\phi(t)dt$ where $\phi(t)$ is integrable, in order to be able to apply Gronwall's inequality.

We first establish this for the contribution from $\ln \mathbf{y}$. Frequency-localize the equation (1.1), resulting in

$$\Box P_k \ln \mathbf{y} = -P_k[\frac{\partial_\nu \mathbf{x}}{\mathbf{y}}\frac{\partial^\nu \mathbf{x}}{\mathbf{y}}] \tag{1.3}$$

[4]We are being imprecise here. All that matters to us is the global bound as stated.

We observe that $\frac{\partial_\nu \mathbf{x}}{\mathbf{y}}\frac{\partial^\nu \mathbf{y}}{\mathbf{y}} = \frac{(\partial_r-\partial_t)\mathbf{x}}{\mathbf{y}}\frac{(\partial_r+\partial_t)\mathbf{x}}{\mathbf{y}}$, and apply a Littlewood-Paley trichotmomy:

$$\begin{aligned}\Box P_k \ln \mathbf{y} = &-P_k[P_{<k-10}\frac{(\partial_r-\partial_t)\mathbf{x}}{\mathbf{y}}P_{[k-5,k+5]}\frac{(\partial_r+\partial_t)\mathbf{x}}{\mathbf{y}}] \\ &-P_k[P_{[k-10,k+10]}\frac{(\partial_r-\partial_t)\mathbf{x}}{\mathbf{y}}P_{<k+15}\frac{(\partial_r+\partial_t)\mathbf{x}}{\mathbf{y}}] \\ &-P_k[P_{>k+10}\frac{(\partial_r-\partial_t)\mathbf{x}}{\mathbf{y}}P_{>k+5}\frac{(\partial_r+\partial_t)\mathbf{x}}{\mathbf{y}}]\end{aligned} \tag{1.4}$$

We use the following terminology: $P_{<a} = \sum_{k<a} P_k$, $a$, $k \in \mathbf{Z}$, $P_{[a,b]} = \sum_{k\in[a,b]} P_k$ etc. We restrict ourselves to time interval $[0,T]$, and let $T \to \infty$. The general case will follow from time reversal symmetry. Applying Duhamel's formula, we see that we need to control the norm $\sum_{k\in\mathbf{Z}} 2^{\delta|k|}||P_k(.)||_{L_t^1L_x^2}$ of the right hand side by an expression $1+\int_0^T A(t)\phi(t)dt$. We may restrict ourselves to a time interval $[c,\infty]$ for some $c>0$ (depending on the initial data), on account of local-in-time well-posedness and finite propagation speed. We estimate each of the terms on the right-hand side of (1.4): *first assume* $k \geq 0$.

(i) *The first term*: We would like to place the 2nd input $P_{[k-5,k+5]}\frac{(\partial_r+\partial_t)\mathbf{x}}{\mathbf{y}}$ into $L_t^\infty L_x^2$ and the first input $P_{<k-10}\frac{(\partial_r-\partial_t)\mathbf{x}}{\mathbf{y}}$ into $L_t^1L_x^\infty$. This doesn't integrate up, however. Placing the 2nd input into $L_t^\infty L_x^\infty$ will work provided $t$ is much larger than $2^{|k|}$, but not in the opposite case: Thus we subdivide

$$\begin{aligned}&P_k[P_{<k-10}\frac{(\partial_r-\partial_t)\mathbf{x}}{\mathbf{y}}P_{[k-5,k+5]}\frac{(\partial_r+\partial_t)\mathbf{x}}{\mathbf{y}}] \\ &= \phi_{\geq 2^{\frac{k}{C}}}(t))P_k[P_{<k-10}\frac{(\partial_r-\partial_t)\mathbf{x}}{\mathbf{y}}P_{[k-5,k+5]}\frac{(\partial_r+\partial_t)\mathbf{x}}{\mathbf{y}}] \\ &+\phi_{<2^{\frac{k}{C}}}(t)P_k[P_{<k-10}\frac{(\partial_r-\partial_t)\mathbf{x}}{\mathbf{y}}P_{[k-5,k+5]}\frac{(\partial_r+\partial_t)\mathbf{x}}{\mathbf{y}}],\end{aligned}$$

where $\phi_{\geq a}(t)$, $\phi_{<a}(t)$ are smooth cutoffs to dilates of the regions $t \geq a$, $t < a$, adding up to 1. Also, $C$ is a large number to be chosen. We can immediately estimate

$$\begin{aligned}&2^{\delta k}||\phi_{\geq 2^{\frac{k}{C}}}(t))P_k[P_{<k-10}\frac{(\partial_r-\partial_t)\mathbf{x}}{\mathbf{y}}P_{[k-5,k+5]}\frac{(\partial_r+\partial_t)\mathbf{x}}{\mathbf{y}}]||_{L_t^1L_x^2} \\ &\leq C2^{\delta k}||P_{<k-10}\frac{(\partial_r-\partial_t)\mathbf{x}}{\mathbf{y}}||_{L_t^\infty L_x^2}||\phi_{\geq 2^{\frac{k}{C}}}(t))P_{[k-5,k+5]}\frac{(\partial_r+\partial_t)\mathbf{x}}{\mathbf{y}}||_{L_t^1L_x^\infty} \\ &\leq C2^{\delta k}2^{-\frac{k}{2C}}\end{aligned}$$

This can be summed over $k \geq 0$ provided $\delta < \frac{1}{2C}$.
Now we proceed to the case in which time is dominated by frequency, $t \leq C2^{\frac{k}{C}}$. We shall distinguish between the region separated from the light cone, where we use lemma 1.7, as well as the region very close to the light cone, where we use

Proposition 1.1 as well as Hoelder's inequality: we decompose

$$P_{[k-5,k+5]}\frac{(\partial_r+\partial_t)\mathbf{x}}{\mathbf{y}} = P_{[k-5,k+5]}(\psi_{>\frac{t}{2}}(|t|-|x|)\frac{(\partial_r+\partial_t)\mathbf{x}}{\mathbf{y}})$$
$$+ P_{[k-5,k+5]}(\psi_{\frac{t}{2}>.\geq 2^{-\mu k}}(|t|-|x|)\frac{(\partial_r+\partial_t)\mathbf{x}}{\mathbf{y}})$$
$$+ P_{[k-5,k+5]}(\psi_{<2^{-\mu k}}(|t|-|x|)\frac{(\partial_r+\partial_t)\mathbf{x}}{\mathbf{y}})$$

where the smooth cutoffs $\psi_{>\frac{t}{2}}(.)$, $\psi_{\frac{t}{2}>.\geq 2^{-\mu k}}(.)$, $\psi_{<2^{-\mu k}}(.)$ add up to 1 and localize, respectively, to dilates of the regions indicated in their subscripts. We let $\mu$ be a small positive number to be chosen. We have

$$2^{\delta k}||\phi_{<2^{\frac{k}{C}}}(t)P_k[P_{<k-10}\frac{(\partial_r-\partial_t)\mathbf{x}}{\mathbf{y}}$$
$$P_{[k-5,k+5]}(\psi_{>\frac{t}{2}}(|t|-|x|)\frac{(\partial_r+\partial_t)\mathbf{x}}{\mathbf{y}})]||_{L_t^1L_x^2([c,T]\times\mathbf{R}^2)}$$
$$\leq C2^{(\delta-1)k}\int_c^T ||P_{<k-10}\frac{(\partial_r-\partial_t)\mathbf{x}}{\mathbf{y}}(t)||_{L_x^\infty}$$
$$||P_{[k-5,k+5]}\nabla_x(\psi_{>\frac{t}{2}}(|t|-|x|)\frac{(\partial_r+\partial_t)\mathbf{x}}{\mathbf{y}})||_{L_x^2}dt$$

Now using lemma 1.7 as well as Proposition 1.1

$$||P_{[k-5,k+5]}\nabla_x(\psi_{>\frac{t}{2}}(|t|-|x|)\frac{(\partial_r+\partial_t)\mathbf{x}}{\mathbf{y}})||_{L_x^2} \leq Ct^{-(2-\epsilon)}\sqrt{\int_0^{\frac{t}{2}} r^{-1-2\epsilon}rdr} \leq Ct^{-\frac{3}{2}}.$$

This implies, reiterating application of Proposition 1.1

$$2^{\delta k}||\phi_{<2^{\frac{k}{C}}}(t)P_k[P_{<k-10}\frac{(\partial_r-\partial_t)\mathbf{x}}{\mathbf{y}}$$
$$P_{[k-5,k+5]}(\psi_{>\frac{t}{2}}(|t|-|x|)\frac{(\partial_r+\partial_t)\mathbf{x}}{\mathbf{y}})]||_{L_t^1L_x^2([c,T]\times\mathbf{R}^2)}$$
$$\leq C2^{(\delta-1)k}\int_c^T t^{-\frac{3}{2}}dt,$$

which can be summed over $k \geq 0$, provided $\delta < 1$. Next, we estimate

$$2^{\delta k}||\phi_{<2^{\frac{k}{C}}}(t)P_k[P_{<k-10}\frac{(\partial_r-\partial_t)\mathbf{x}}{\mathbf{y}}$$
$$P_{[k-5,k+5]}(\psi_{\frac{t}{2}>.\geq 2^{-\mu k}}(|t|-|x|)\frac{(\partial_r+\partial_t)\mathbf{x}}{\mathbf{y}})]||_{L_t^1L_x^2([c,T]\times\mathbf{R}^2)}$$
$$\leq C2^{(\delta-1)k}||P_{<k-10}\frac{(\partial_r-\partial_t)\mathbf{x}}{\mathbf{y}}||_{L_t^\infty L_x^2}$$
$$||P_{[k-5,k+5]}\nabla_x(\psi_{\frac{t}{2}>.\geq 2^{-\mu k}}(|t|-|x|)\frac{(\partial_r+\partial_t)\mathbf{x}}{\mathbf{y}})||_{L_t^1L_x^\infty([c,T]\times\mathbf{R}^2)}$$
$$\leq C2^{(\delta-1+\mu)k}||P_{<k-10}\frac{(\partial_r-\partial_t)\mathbf{x}}{\mathbf{y}}||_{L_t^\infty L_x^2}\int_c^T t^{-\frac{3}{2}}dt.$$

We can sum here over $k$ provided $\mu+\delta<1$. Finally, we calculate using Hoelder's inequality as well as Proposition 1.1

$$\begin{aligned}
&2^{\delta k}||\phi_{<2^{\frac{k}{C}}}(t)P_k[P_{<k-10}\frac{(\partial_r-\partial_t)\mathbf{x}}{\mathbf{y}}\\
&\qquad P_{[k-5,k+5]}(\psi_{<2^{-\mu k}}(|t|-|x|)\frac{(\partial_r+\partial_t)\mathbf{x}}{\mathbf{y}})||_{L_t^1L_x^2([c,T]\times\mathbf{R}^2)}\\
&\leq C\min\{T,2^{\frac{k}{C}}\}2^{\delta k}||P_{<k-10}\frac{(\partial_r-\partial_t)\mathbf{x}}{\mathbf{y}}||_{L_x^\infty}||P_{[k-5,k+5]}(\psi_{<2^{-\mu k}}(|t|-|x|)\frac{(\partial_r+\partial_t)\mathbf{x}}{\mathbf{y}})||_{L_x^2}\\
&\leq C2^{(\delta+\frac{1}{C}-\frac{\mu}{2})k}
\end{aligned}$$

This can be summed over $k\geq 0$ provided we have $\delta+\frac{1}{C}<\frac{\mu}{2}$. Combining with the conditions obtained earlier, namely $\delta<\frac{1}{2C}$ as well as $\delta+\mu<1$, we get $\delta<\frac{1}{7}$.

(ii) *The 2nd term of* (1.4) . This term appears immediate on account of Proposition 1.1. Formally

$$2^{\delta k}||P_k[P_{[k-10,k+10]}\frac{(\partial_r-\partial_t)\mathbf{x}}{\mathbf{y}}P_{<k+15}\frac{(\partial_r+\partial_t)\mathbf{x}}{\mathbf{y}}]||_{L_t^1L_x^2([c,T]\times\mathbf{R}^2)}\leq C\int_c^T A(t)(1+t)^{-\frac{3}{2}}dt$$

We have to argue more carefully here since $\partial_u$ involves $\partial_r=\frac{x_1}{r}\partial_{x_1}+\frac{x_2}{r}\partial_{x_2}$. Decompose

$$\begin{aligned}
P_k(\frac{\partial_r\mathbf{x}}{\mathbf{y}})&=P_k[P_{<k-10}(\frac{x_1}{r})P_{[k-10,k+10]}(\frac{\partial_1\mathbf{x}}{\mathbf{y}})\\
&+P_{[k-10,k+10]}(\frac{x_1}{r})P_{<k+15}(\frac{\partial_1\mathbf{x}}{\mathbf{y}})+P_{>k+10}(\frac{x_1}{r})P_{>k+5}(\frac{\partial_1\mathbf{x}}{\mathbf{y}})]\\
&+\text{similar terms}
\end{aligned}\tag{1.5}$$

The first term in this Littlewood-Paley trichotomy is estimated exactly as before, so we treat the 2nd and third term. Let $\chi_0(.)\in C_0^\infty(\mathbf{R}_{>0})$ be the cutoff used for the Littlewood-Paley localizers $P_k$. We note that

$$P_k(\frac{x_1}{r})(x)=2^{2k}\int_{\mathbf{R}^2}\widehat{\chi_0}(2^k(x-y))\frac{y_1}{|y|}dy=2^{2k}\int_{\mathbf{R}^2}\widehat{\chi_0}(2^ky)[\frac{y_1-x_1}{|y-x|}-\frac{x_1}{|x|}]dy$$

On account of the inequality

$$|\frac{y_1-x_1}{|y-x|}-\frac{x_1}{|x|}|\leq C\min\{\frac{|y|}{|x|},1\},$$

we get, using the rapid decay of $y\to\widehat{\chi_0}(2^ky)$ outside of a disc of radius $\sim 2^{-k}$:

$$|P_k(\frac{x_1}{r})(x)|\leq C\min\{\frac{2^{-k}}{|x|},1\}.$$

We introduce another cutoff $\chi_{2^{-\frac{k}{2}}}(x)$ which smoothly localizes to a dilate of the disc $B_{2^{-\frac{k}{2}}}(\mathbf{0})$ centered at $\mathbf{0} = (0,0)$. We then decompose

$$P_k[P_{[k-10,k+10]}(\frac{x_1}{r})P_{<k+15}(\frac{\partial_1 \mathbf{x}}{\mathbf{y}})P_{<k+15}\frac{\partial_v \mathbf{x}}{\mathbf{y}}] =$$
$$P_k[\chi_{2^{-\frac{k}{2}}}(x)P_{[k-10,k+10]}(\frac{x_1}{r})P_{<k+15}(\frac{\partial_1 \mathbf{x}}{\mathbf{y}})P_{<k+15}\frac{\partial_v \mathbf{x}}{\mathbf{y}}]$$
$$+ P_k[(1-\chi_{2^{-\frac{k}{2}}}(x))P_{[k-10,k+10]}(\frac{x_1}{r})P_{<k+15}(\frac{\partial_1 \mathbf{x}}{\mathbf{y}})P_{<k+15}\frac{\partial_v \mathbf{x}}{\mathbf{y}}]$$

Using Proposition 1.1 and Hoelder's inequality, we get

$$2^{\delta k}||P_k[\chi_{2^{-\frac{k}{2}}}(x)P_{[k-10,k+10]}(\frac{x_1}{r})P_{<k+15}(\frac{\partial_1 \mathbf{x}}{\mathbf{y}})P_{<k+15}\frac{\partial_v \mathbf{x}}{\mathbf{y}}]||_{L_1^1 L_x^2}$$
$$\leq C2^{\delta k}||\chi_{2^{-\frac{k}{2}}}(x)P_{[k-10,k+10]}(\frac{x_1}{r})||_{L_t^\infty L_x^2}||P_{<k+15}(\frac{\partial_1 \mathbf{x}}{\mathbf{y}})P_{<k+15}\frac{\partial_v \mathbf{x}}{\mathbf{y}}||_{L_t^1 L_x^\infty}$$
$$\leq C2^{(\delta-\frac{1}{2})k}.$$

On the other hand, using the preceding calculations as well as Proposition 1.1 we get

$$2^{\delta k}||P_k[(1-\chi_{2^{-\frac{k}{2}}}(x))P_{[k-10,k+10]}(\frac{x_1}{r})P_{<k+15}(\frac{\partial_1 \mathbf{x}}{\mathbf{y}})P_{<k+15}\frac{\partial_v \mathbf{x}}{\mathbf{y}}]||_{L_t^1 L_x^2}$$
$$\leq C2^{\delta k}||(1-\chi_{2^{-\frac{k}{2}}}(x))P_{[k-10,k+10]}(\frac{x_1}{r})||_{L_t^\infty L_x^\infty}$$
$$||P_{<k+15}(\frac{\partial_1 \mathbf{x}}{\mathbf{y}})||_{L_t^\infty L_x^2}||P_{<k+15}\frac{\partial_v \mathbf{x}}{\mathbf{y}}||_{L_t^1 L_x^\infty}$$
$$\leq C2^{(\delta-\frac{1}{2})k}.$$

Since we have to choose $\delta < \frac{1}{7}$, both can be summed over $k \geq 0$. The case corresponding to the third term in (1.5) as well as the remaining terms are handled analogously.

(iii) *The third term of* (1.4): We can write

$$P_k[P_{>k+10}(\frac{\partial_u \mathbf{x}}{\mathbf{y}})P_{>k+5}(\frac{\partial_v \mathbf{x}}{\mathbf{y}})] = \sum_{l_1>k+10,\,|l_1-l_2|<5} P_k[P_{l_1}(\frac{\partial_u \mathbf{x}}{\mathbf{y}})P_{l_2}(\frac{\partial_v \mathbf{x}}{\mathbf{y}})]$$

Next, we estimate, using Proposition 1.1

$$\sum_{l_1>k+10,\,|l_1-l_2|<5} 2^{\delta k}||P_k[P_{l_1}(\frac{\partial_u \mathbf{x}}{\mathbf{y}})P_{l_2}(\frac{\partial_v \mathbf{x}}{\mathbf{y}})]||_{L_t^1 L_x^2([c,T]\times\mathbf{R}^2)}$$
$$\leq C\sum_{l_1>k+10,\,|l_1-l_2|<5} 2^{\delta(k-l_1)}\int_c^T [2^{\delta l_1}||P_{l_1}(\frac{\partial_u \mathbf{x}}{\mathbf{y}})(t,.)||_{L_x^2}]t^{-\frac{3}{2}}dt,$$

and one can sum here over both $l_1$, $k$ to obtain the upper bound $\leq C1+\int_c^T A(t)t^{-\frac{3}{2}}dt$. We have to argue for $\frac{\partial_u \mathbf{x}}{\mathbf{y}}$ as in (ii). This completes the estimates for case $k \geq 0$. For the case $k < 0$, we have for $M > 6$

$$||P_k[\frac{\partial_u \mathbf{x}}{\mathbf{y}}\frac{\partial_v \mathbf{x}}{\mathbf{y}}]||_{L_t^1 L_x^2} \leq C2^{\frac{2k}{M}}||\frac{\partial_u \mathbf{x}}{\mathbf{y}}||_{L_t^\infty L_x^2}||\frac{\partial_v \mathbf{x}}{\mathbf{y}}||_{L_t^1 L_x^M} \leq C2^{\frac{2k}{M}},$$

We have used *Bernstein's inequality* which states that for any rectangle $R \subset \mathbf{R}^2$ and smooth cutoff $\chi_R$ supported in $R$ we have[5]

$$||\mathcal{F}^{-1}(\chi_R \mathcal{F}f)||_{L^q_x} \leq C|R|^{\frac{1}{p}-\frac{1}{q}}||f||_{L^p_x},\, p \leq q.$$

Also, the the estimate for $||\frac{\partial_\nu \mathbf{x}}{\mathbf{y}}||_{L^1_t L^M_x}$ follows from interpolating between the decay estimate for $||\frac{\partial_\nu \mathbf{x}}{\mathbf{y}}(t)||_{L^\infty_x}$ and energy conservation.

The estimates for $\frac{\partial_\nu \mathbf{x}}{\mathbf{y}}$ are similar: we have by the same reasoning as before

$$\sum_{k\in\mathbf{Z}} 2^{\delta k}||P_k[\frac{\partial_\nu \mathbf{y}\partial^\nu \mathbf{y} + \partial_\nu \mathbf{x}\partial^\nu \mathbf{x}}{\mathbf{y}^2}]||_{L^1_t L^2_x([c,T]\times\mathbf{R}^2)} \leq C1 + \int_c^T A(t)t^{-\frac{3}{2}}dt$$

Now as for the nonlinearity on the right hand side of (1.2), the small frequency case $k < 0$ follows exactly as above from the boundedness of $\frac{\mathbf{x}}{\mathbf{y}}$, see lemma 1.6. As for the large frequency case, we have the usual frequency trichotomy

$$\begin{aligned} P_k[\frac{\mathbf{x}}{\mathbf{y}}\frac{\partial_\nu \mathbf{y}\partial^\nu \mathbf{y} + \partial_\nu \mathbf{x}\partial^\nu \mathbf{x}}{\mathbf{y}^2}] &= P_k[P_{[k-5,k+5]}(\frac{\mathbf{x}}{\mathbf{y}})P_{<k-10}[\frac{\partial_\nu \mathbf{y}\partial^\nu \mathbf{y} + \partial_\nu \mathbf{x}\partial^\nu \mathbf{x}}{\mathbf{y}^2}]] \\ &+ P_k[P_{<k+15}(\frac{\mathbf{x}}{\mathbf{y}})P_{[k-10,k+10]}[\frac{\partial_\nu \mathbf{y}\partial^\nu \mathbf{y} + \partial_\nu \mathbf{x}\partial^\nu \mathbf{x}}{\mathbf{y}^2}]] \\ &+ P_k[P_{>k+5}(\frac{\mathbf{x}}{\mathbf{y}})P_{>k+10}[\frac{\partial_\nu \mathbf{y}\partial^\nu \mathbf{y} + \partial_\nu \mathbf{x}\partial^\nu \mathbf{x}}{\mathbf{y}^2}]] \end{aligned}$$

We need

LEMMA 1.8. *The following inequality holds:*

$$\sum_{k\geq 0} 2^{\delta k}||P_k\nabla_x(\frac{\mathbf{x}}{\mathbf{y}})(t)||_{L^2_x} \leq CA(t) + 1$$

PROOF. : Call the left hand side $B(t)$. Note that

$$B(t) \leq A(t) + \sum_{k\geq 0} 2^{\delta k}||P_k(\frac{\mathbf{x}}{\mathbf{y}}\frac{\nabla \mathbf{y}}{\mathbf{y}})(t)||_{L^2_x}$$

We have the frequency trichotomy

$$\begin{aligned} P_k[\frac{\nabla \mathbf{y}}{\mathbf{y}}\frac{\mathbf{x}}{\mathbf{y}}] &= P_k[P_{<k-10}(\frac{\nabla \mathbf{y}}{\mathbf{y}})P_{[k-5,k+5]}(\frac{\mathbf{x}}{\mathbf{y}})] \\ &+ P_k[P_{[k-10,k+10]}(\frac{\nabla \mathbf{y}}{\mathbf{y}})P_{<k+15}(\frac{\mathbf{x}}{\mathbf{y}})] + P_k[P_{>k+10}(\frac{\nabla \mathbf{y}}{\mathbf{y}})P_{>k+5}(\frac{\mathbf{x}}{\mathbf{y}})] \end{aligned}$$

The estimate is immediate for the 2nd term on the right hand side. As to the first, we have

$$\begin{aligned} &2^{\delta k}||P_k[P_{<k-10}(\frac{\nabla \mathbf{y}}{\mathbf{y}})P_{[k-5,k+5]}(\frac{\mathbf{x}}{\mathbf{y}})](t)||_{L^2_x} \\ &\leq C2^{(\delta-1)k}||P_{<k-10}(\frac{\nabla \mathbf{y}}{\mathbf{y}})||_{L^\infty_x}||\nabla_x P_{[k-5,k+5]}(\frac{\mathbf{x}}{\mathbf{y}})(t)||_{L^2_x} \\ &\leq C2^{(\delta-1)k}B(t) \end{aligned}$$

[5]We denote the spatial Fourier transform of $f(x)$ either by $\hat{f}$ or $\mathcal{F}f$.

One can also estimate this term by $\leq C2^{\delta k}$ from energy conservation and lemma 1.6. The estimate for the third term in the preceding trichotomy is similar. We conclude that

$$B(t) \leq CA(t) + \sum_{0\leq k\leq C} 2^{\delta k} + \sum_{k>C} 2^{(\delta-1)k} B(t)$$

Choosing $C$ large enough, one obtains the claim of the lemma. □

Armed with this, we now have (we may assume $k \geq 10$)

$$2^{\delta k}||P_k[P_{[k-5,k+5]}(\frac{\mathbf{x}}{\mathbf{y}})P_{<k-10}[\frac{\partial_\nu\mathbf{y}\partial^\nu\mathbf{y}+\partial_\nu\mathbf{x}\partial^\nu\mathbf{x}}{\mathbf{y}^2}]]||_{L_t^1L_x^2([c,T]\times\mathbf{R}^2)}$$
$$\leq C2^{(\delta-1)k}\int_c^T ||\nabla_x P_{[k-5,k+5]}(\frac{\mathbf{x}}{\mathbf{y}})(t)||_{L_x^2}||P_{<k-10}[\frac{\partial_\nu\mathbf{y}\partial^\nu\mathbf{y}+\partial_\nu\mathbf{x}\partial^\nu\mathbf{x}}{\mathbf{y}^2}(t)||_{L_x^\infty}dt$$

Using Proposition 1.1 as well as the preceding lemma and summing over $k \geq 10$, we bound this by $\leq C1 + \int_c^T A(t)t^{-\frac{3}{2}}dt$. The estimate for the third term in the frequency trichotomy preceding the last lemma is more of the same. Thus we get

$$\sum_{k\in\mathbf{Z}} 2^{\delta|k|}||P_k[\frac{\mathbf{x}}{\mathbf{y}}\frac{\partial_\nu\mathbf{y}\partial^\nu\mathbf{y}+\partial_\nu\mathbf{x}\partial^\nu\mathbf{x}}{\mathbf{y}^2}]||_{L_t^1L_x^2([c,T]\times\mathbf{R}^2)} \leq C1 + \int_c^T A(t)t^{-\frac{3}{2}}dt$$

Using Duhamel's formula, we get

$$\sum_{k\in\mathbf{Z}} 2^{\delta|k|}[||P_k\nabla_x(\frac{\mathbf{x}}{\mathbf{y}})(T)||_{L_x^2} + ||P_k\partial_t(\frac{\mathbf{x}}{\mathbf{y}})(T)||_{L_x^2}] \leq C1 + \int_c^T A(t)t^{-\frac{3}{2}}dt$$

We need to estimate $\frac{\nabla\mathbf{x}}{\mathbf{y}}(T)$, which differs from the preceding by $\frac{\mathbf{x}}{\mathbf{y}}\frac{\nabla\mathbf{y}}{\mathbf{y}}(T)$. For frequencies $\geq 0$, this is estimated as in the preceding lemma, observing that we already improved the estimate for $||\frac{\nabla\mathbf{y}}{\mathbf{y}}(T)||_{L_x^2}$ from the preceding estimates (i)-(iii). The only case not yet covered concerns small frequencies. However, we have for $k < 0$

$$P_k[\frac{\mathbf{x}}{\mathbf{y}}\frac{\nabla\mathbf{y}}{\mathbf{y}}] = P_k[P_{[k-5,k+5]}(\frac{\mathbf{x}}{\mathbf{y}})P_{<k-10}(\frac{\nabla\mathbf{y}}{\mathbf{y}})]$$
$$+ P_k[P_{<k+15}(\frac{\mathbf{x}}{\mathbf{y}})P_{[k-10,k+10]}(\frac{\nabla\mathbf{y}}{\mathbf{y}})] + P_k[P_{>k+5}(\frac{\mathbf{x}}{\mathbf{y}})P_{>k+10}(\frac{\nabla\mathbf{y}}{\mathbf{y}})]$$

Then

$$\sum_{k<0} 2^{\delta|k|}||P_k[P_{[k-5,k+5]}(\frac{\mathbf{x}}{\mathbf{y}})P_{<k-10}(\frac{\nabla\mathbf{y}}{\mathbf{y}})](T)||_{L_x^2}$$
$$\leq C\sum_{k<0} 2^{\delta|k|}||P_{[k-5,k+5]}(\frac{\mathbf{x}}{\mathbf{y}})||_{L_x^\infty}||P_{<k-10}(\frac{\nabla\mathbf{y}}{\mathbf{y}})(T)||_{L_x^2},$$

which in turn is bounded by $\leq C1 + \int_c^T A(t)t^{-\frac{3}{2}}dt$, as is easily[6] verified. The estimate for the 2nd term is immediate and the estimate for the third term as

[6]Use Bernstein's inequality.

follows:

$$\sum_{k\in\mathbf{Z}_{<0}} 2^{\delta|k|}||P_k[P_{>k+5}(\frac{\mathbf{x}}{\mathbf{y}})P_{>k+10}(\frac{\nabla\mathbf{y}}{\mathbf{y}})](T)||_{L^2_x}$$
$$\leq C\sum_{k\in\mathbf{Z}_{<0}}\sum_{l_1>k+10,\,|l_1-l_2|<5}||P_k[P_{l_2}(\frac{\mathbf{x}}{\mathbf{y}})P_{l_1}(\frac{\nabla\mathbf{y}}{\mathbf{y}})](T)||_{L^2_x}$$
$$\leq C\sum_{k\in\mathbf{Z}_{<0}}\sum_{l_1>k+10,\,|l_1-l_2|<5} 2^{\delta|k|}2^{k-l_1}||P_{l_2}\nabla_x(\frac{\mathbf{x}}{\mathbf{y}})(T)||_{L^2_x}||P_{l_1}(\frac{\nabla\mathbf{y}}{\mathbf{y}})||_{L^2_x}$$

One verifies easily from the preceding estimates that this is $\leq C1+\int_c^\infty A(t)t^{-\frac{3}{2}}dt$. Putting all of these ingredients together, we obtain

$$A(t)\leq C1+\int_c^T A(t)t^{-\frac{3}{2}}dt.$$

The desired upper bound now follows from Gronwall's inequality. □

COROLLARY 1.1. *Let $N(\nabla\mathbf{x},\,\nabla\mathbf{y},\,\mathbf{x},\,\mathbf{y})$ denote any one of the nonlinearities occuring on the right hand side of* (1.1), (1.2). *Then for $\delta$ as in Proposition 1.2, we have for $\delta<\frac{1}{7}$*

$$\sum_{k\in\mathbf{Z}} 2^{\delta|k|}||P_kN(\nabla\mathbf{x},\,\nabla\mathbf{y},\,\mathbf{x},\,\mathbf{y})||_{L^1_tL^2_x([-T,T]\times\mathbf{R}^2)}<\infty$$

This follows from the preceding proof and time reversal symmetry. In the same vein, we have the following lemma:

LEMMA 1.9. *Choosing $\delta>0$ small enough, we have the inequality*

$$\sum_{k\in\mathbf{Z}} 2^{\delta|k|}||P_kN(\nabla\mathbf{x},\,\nabla\mathbf{y},\,\mathbf{x},\,\mathbf{y})||_{L^2_t\dot{H}^{-\frac{1}{2}}}<\infty$$

PROOF. : We work with $N(...)=\frac{\partial_\nu\mathbf{x}}{\mathbf{y}}\frac{\partial^\nu\mathbf{x}}{\mathbf{y}}$, the other cases being similar. Divide into the cases $k\geq 0$ and $k<0$. In the first case, estimate

$$2^{\delta k}||P_kN(\nabla\mathbf{x},\,\nabla\mathbf{y},\,\mathbf{x},\,\mathbf{y})||_{L^2_t\dot{H}^{-\frac{1}{2}}}$$
$$\leq C2^{(\delta-\frac{1}{2})k}||\frac{(\partial_t-\partial_r)\mathbf{x}}{\mathbf{y}}||_{L^\infty_tL^2_x}||\frac{(\partial_t+\partial_r)\mathbf{x}}{\mathbf{y}}||_{L^2_tL^\infty_x}\leq C2^{(\delta-\frac{1}{2})k}$$

In the 2nd case, estimate

$$2^{-\delta k}||P_kN(\nabla\mathbf{x},\,\nabla\mathbf{y},\,\mathbf{x},\,\mathbf{y})||_{L^2_t\dot{H}^{-\frac{1}{2}}}$$
$$\leq C2^{(-\delta+\frac{2}{4-}-\frac{1}{2})k}||\frac{(\partial_t-\partial_r)\mathbf{x}}{\mathbf{y}}||_{L^\infty_tL^2_x}||\frac{(\partial_t+\partial_r)\mathbf{x}}{\mathbf{y}}||_{L^2_tL^{4-}_x}\leq C2^{(-\delta+\frac{2}{4-}-\frac{1}{2})k}$$

We have used here that

$$||\frac{(\partial_t+\partial_r)\mathbf{x}}{\mathbf{y}}(t)||_{L^{4-}_x}\leq Ct^{-\frac{3}{4+}}$$

which follows from interpolating between Proposition 1.1 and energy conservation. Choosing $\delta<\frac{2}{4-}-\frac{1}{2}$ results in the claim of the lemma. □

## 1.3. The perturbation argument

**1.3.1. Precise statement of theorem. Outline of the procedure.** The formulation (1.1), (1.2), while good enough for the purposes of the last section, will not suffice for us here[7]. Instead, following the procedure in [**23**], we shall pass to the derivative formulation of the problem, and translate everything into the Coulomb Gauge. More precisely, introduce the variables $\phi_\nu^1 = -\frac{\partial_\nu \mathbf{x}}{\mathbf{y}}$, $\phi_\nu^2 = -\frac{\partial_\nu \mathbf{y}}{\mathbf{y}}$, pass to complex notation $\phi_\nu = \phi_\nu^1 + i\phi_\nu^2$, and revert to the Coulomb Gauge by introducing the variables $\psi_\nu = \phi_\nu e^{-i\triangle^{-1}\sum_{i=1,2}\partial_i\phi_i^1}$. One gets the following remarkable self-contained divergence curl system:

(1.6)

$$\partial_\alpha\psi_\beta - \partial_\beta\psi_\alpha = i\psi_\beta\triangle^{-1}\sum_{j=1,2}\partial_j(\psi_\alpha^1\psi_j^2 - \psi_\alpha^2\psi_j^1) - i\psi_\alpha\triangle^{-1}\sum_{j=1,2}\partial_j(\psi_\beta^1\psi_j^2 - \psi_\beta^2\psi_j^1)$$

$$\partial_\nu\psi^\nu = i\psi^\nu\triangle^{-1}\sum_{j=1,2}\partial_j(\psi_\nu^1\psi_j^2 - \psi_\nu^2\psi_j^1). \tag{1.7}$$

From these one easily deduces the following system of wave equations:

$$\begin{aligned}\Box\psi_\alpha =& i\partial^\beta[\psi_\alpha\triangle^{-1}\sum_{j=1}^{2}\partial_j[\psi_\beta^1\psi_j^2 - \psi_\beta^2\psi_j^1]]\\ &- i\partial^\beta[\psi_\beta\triangle^{-1}\sum_{j=1}^{2}\partial_j[\psi_\alpha^1\psi_j^2 - \psi_\alpha^2\psi_j^1]]\\ &+ i\partial_\alpha[\psi_\nu\triangle^{-1}\sum_{j=1}^{2}\partial_j[\psi^{1\nu}\psi_j^2 - \psi^{2\nu}\psi_j^1]].\end{aligned} \tag{1.8}$$

As in [**23**], these in conjunction with the underlying first-order system (1.6), (1.7) shall form the basis for our estimates. We can now give the precise version of Theorem 1.3:

THEOREM 1.10. *Let $u : \mathbf{R}^{2+1} \to \mathbf{H}^2$ be a smooth spherically symmetric Wave Map with compactly supported (large) initial data. Let $\{\psi_\nu\}_{\nu=0}^2$ be the derivative components in the Coulomb Gauge. Then for any $\mu > 0$ there exists $\epsilon = \epsilon(u,\mu) > 0$ such that for all smooth initial data $\tilde{u}[0] = (\tilde{u}(0), \partial_t\tilde{u}(0))$ with $||(u - \tilde{u})[0]||_{H^{1+\mu}\times H^\mu} < \epsilon$, there exists a smooth Wave Map $\tilde{u}$ extending $\tilde{u}[0]$. Also, $\tilde{u}$ will stay close to $u$ in the sense that*

$$\sup_t ||(\psi_\nu - \tilde{\psi_\nu})(t)||_{L_x^2} \leq C\epsilon$$

[7]It appears that the fact that we impose stronger control over $\tilde{u}$ than just the energy (indeed stronger than a Besov norm) should allow us to work with the original coordinate formulation, see e. g. [**40**]. However, it appears that the bilinear null-structure in (1.1), (1.2) is not good enough to obtain the gains in time we shall need, see Proposition 2.2. Indeed, proving an equivalent of this Proposition for the bilinear expressions appears to require time decay (in the sense that the norm evaluated on the function truncated to large times decays) for norms such as $||u||_{\dot{X}_k^{1,\frac{1}{2},\infty}}$, which already fails for free waves. Moreover, our proof will actually reveal that one gets an honest $H^1$-stability result provided one restricts oneself to large enough times.

The proof of this shall consist in analyzing the wave equation satisfied by the difference $\delta\psi_\nu := \tilde{\psi}_\nu - \psi_\nu$. Subtracting the wave equations for $\tilde{\psi}_\nu$, $\psi_\nu$, and eliminating the $\tilde{\psi}_\nu$ results in terms linear, quadratic and cubic in $\delta\psi_\nu$. As these expressions have no apparent null-structure in them, we shall revert to the device of a Hodge-type decomposition used already in [**22**], [**23**]: we shall write $\psi_\nu = R_\nu\psi + \chi_\nu$ and similarly for $\tilde{\psi}_\nu$, where we impose the condition $\sum_{i=1,2}\partial_i\chi_i = 0$. Note that this results in a similar decomposition for $\delta\psi_\nu$. One easily deduces an elliptic div-curl system for $\chi_\nu$, $\tilde{\chi}_\nu$, from which one deduces the schematic identities $\chi_\nu = \nabla^{-1}(\psi\nabla^{-1}(\psi^2))$ etc., where the operators $\nabla^{-1}$ stand for linear combinations of operators of the form $\triangle^{-1}\partial_j$. Plugging these ingredients back into the wave equations satisfied by the $\delta\psi_\nu$ and eliminating all $\tilde{\psi}_\nu$ results in trilinear null-form terms as well terms of higher degree of linearity, either linear or of higher degree in the $\delta\psi_\nu$. All of this is just like in [**23**]. Terms which are at least quadratic in the $\delta\psi_\nu$ can be treated just as there, using the fact that Corollary 1.1 shall allow us to retrieve all the necessary estimates about $\psi_\nu$. The only added difficulty comes from the terms linear in $\delta\psi_\nu$. One way to think of these is as an extra driving term added to the flat operator $\Box$. However, the very good decay estimates satisfied by the $\psi_\nu$ shall allow us to treat these terms as source terms instead. The added difficulty over [**23**] we encounter here has to do with the fact that we need to gain explicitly in time. This will force us to localize simultaneously in physical and frequency space. In fact, we shall use a kind of wave packet decomposition to get the necessary estimates. The next two subsections provide the technical setup. In the same vein as the preceding theorem, we have

THEOREM 1.11. *Let $u : \mathbf{R}^{2+1} \to \mathbf{H}^2$ be a smooth geodesic Wave Map with compactly supported initial data. Then for any $\mu > 0$, there exists $\epsilon = \epsilon(u,\mu) > 0$ such that for all smooth initial data $\tilde{u}[0] = (\tilde{u}(0), \partial_t\tilde{u}(0))$, with $||(u-\tilde{u})(0)||_{H^{1+\mu}\times H^{\mu}} < \epsilon$, there exists a smooth Wave Map $\tilde{u} : \mathbf{R}^{2+1} \to \mathbf{H}^2$ extending $\tilde{u}[0]$. $\tilde{u}$ will stay close to $u$ in the sense that $\sup_t ||(\psi_\nu - \tilde{\psi}_\nu)(t)||_{L^2_x} \le C\epsilon$.*

CHAPTER 2

# Technical Preliminaries. Proofs of Main Theorems

**2.0.2. Sobolev type spaces.** We commence by introducing the functional analytic framework of [**40**], [**37**], [**23**] which we have to rely on to run the perturbation argument. We recall the Littlewood-Paley multipliers $P_k$ introduced in the previous section:

$$\widehat{P_k f}(\xi) = \chi_0(\frac{|\xi|}{2^k})\hat{f}(\xi),$$

for a suitable cutoff $\chi_0(.)$. These are not flexible enough, and we also introduce the multipliers $Q_j$ which localize the space-time Fourier support to dyadic distance $\sim 2^j$ from the light cone: letting

$$\tilde{\phi}(\xi,\tau) = \int_{\mathbf{R}^{2+1}} e^{-i(\tau t+\xi\cdot x)}\phi(t,\, x)dtdx$$

denote the space-time Fourier transform, we let

$$\widetilde{Q_j\phi}(\tau,\xi) := \chi_0(\frac{||\tau|-|\xi||}{2^j})\tilde{\phi}(\tau,\xi)$$

where $\chi_0(.)$ is as for the $P_k$'s. We note that these definitions entail the identities

$$\sum_{k\in\mathbf{Z}} P_k\phi = \phi,\, \sum_{j\in\mathbf{Z}} Q_j\phi = \phi,\, \phi\in\mathcal{S}(\mathbf{R}^{2+1}).$$

We have the basic inhomogeneous Sobolev spaces $H^s$, and their homogeneous counterparts $\dot{H}^s$:

$$||\phi||_{H^s} = \sqrt{\int_{\mathbf{R}^2}(1+|\xi|^2)^s|\hat{\phi}(\xi)|^2d\xi},\ ||\phi||_{\dot{H}^s} = \sqrt{\int_{\mathbf{R}^2}|\xi|^{2s}|\hat{\phi}(\xi)|^2d\xi}$$

Note that the space $H^s$ is defined as completion of $\mathcal{S}(\mathbf{R}^2)$ with respect to the first norm. Trying to do the same for $\dot{H}^s$ leads to difficulties (one gets not necessarily locally integrable functions). We shall only work with smooth functions anyways, so we only care about $||.||_{\dot{H}^s}$. These norms are not flexible enough, and we also need the $X^{s,\theta}$ spaces of Klainerman-Machedon as well as their ('frequency localized') homogeneous Besov analogs(again only as norms):

$$||\phi||_{X^{s,\theta}} := \sqrt{\int_{\mathbf{R}^{2+1}}(1+|\xi|^2)^s(1+||\tau|-|\xi||)^{2\theta}|\tilde{\phi}(\tau,\xi)|^2d\tau d\xi},\ \theta > \frac{1}{2}$$

$$||\phi||_{\dot{X}_k^{a,b,c}} := 2^{ka}(\sum_{j\in\mathbf{Z}}[2^{bj}||Q_j\phi||_{L_t^2L_x^2}]^c)^{\frac{1}{c}},\ c<\infty$$

$$||\phi||_{\dot{X}_k^{a,b,\infty}} := 2^{ak}\sup_{j\in\mathbf{Z}}[2^{bj}||Q_j\phi||_{L_t^2L_x^2}]$$

We shall always have $b = \frac{1}{2}$. The latter norms can be assembled to 'global versions', most naturally via

$$||\phi||_{\dot{X}^{a,b,c}} := \sqrt{\sum_{k\in\mathbf{Z}} ||P_k\phi||^2_{\dot{X}_k^{a,b,c}}}$$

The most intuitive way to think about the $X^{s,\theta}$ etc is to view them as superpositions of 'twisted free waves', gotten by foliating space-time by cones $||\tau| - |\xi|| = \lambda$. One has the representation (see [**17**])

$$\phi = \int_{\lambda\in\mathbf{R}} \phi_\lambda e^{it\lambda} d\lambda,$$

where $\Box\phi_\lambda = 0$ and

$$\int_\lambda ||\phi_\lambda||_{H^s} d\lambda \leq C||\phi||_{X^{s,\theta}}.$$

At the homogeneous level, we have the embedding[1]

$$\dot{X}_k^{0,\frac{1}{2},1} \subset L_t^\infty L_x^2.$$

More generally, the Strichartz estimates (see e. g. [**9**]) imply that the following embeddings hold:

$$\dot{X}_k^{0,\frac{1}{2},1} \subset 2^{k(1-\frac{1}{p}-\frac{2}{q})} L_t^p L_x^q,$$

where $\frac{1}{p} + \frac{1}{2q} \leq \frac{1}{4}$. Similar embeddings hold for the 'subcritical spaces' $X^{s,\theta}$. We shall need slightly shrunk versions of the spaces $X^{s,\theta}$ etc. which give stronger control for the 'elliptic regions' far away from the light cone. For example, we have the norms (see [**17**]) $||.||_{\mathcal{X}^{s,\theta}}$, which are defined via

$$||\phi||_{\mathcal{X}^{s,\theta}} := ||\phi||_{X^{s,\theta}} + ||\partial_t\phi||_{X^{s-1,\theta}}$$

Similarly, we introduce the space $\mathcal{H}^s$ defined as the completion of $\mathcal{S}(\mathbf{R}^2)$ under the norm

$$||\psi||_{\mathcal{H}^s} := ||\psi||_{H^s} + ||\partial_t\psi||_{H^{s-1}}$$

**2.0.3. Tataru's null-frame spaces.** This subsection also summarizes material expounded in greater detail elsewhere (e. g. [**37**], [**22**], [**23**]). The spaces $X^{s,\theta}$ and their homogeneous Besov counterparts are unfortunately only part of the story. This has to do with the fact that even the strongest homogeneous versions of these norms (the norms $||.||_{X^{a,\frac{1}{2},1}}$) do not yield good algebra type estimates, due to logarithmic divergences in low frequencies. A solution to this problem is given by 'spaces' incorporating Tataru's null-frame spaces. We present here a first version of spaces that overcome this difficulty. We shall construct norms $||.||_S$ assembled from a family of 'frequency localized' norms $||.||_{S[k]}$:

$$||\phi||_S := \sqrt{\sum_{k\in\mathbf{Z}} ||P_k\phi||^2_{S[k]}}$$

The norms $||.||_{S[k]}$ in turn are gotten as in [**23**]: they are constructed to satisfy $||.||_{\dot{X}_k^{0,\frac{1}{2},\infty}} \leq ||.||_{S[k]} \leq C||.||_{\dot{X}_k^{0,\frac{1}{2},1}}$. We arrange that the norms are invariant under the natural scaling operation associated with derivatives of Wave Maps in $2+1$ dimensions, since we shall be working at the level of the derivative. The precise

[1]The way to think about these is in the sense of inequalities between the associated norms: $A \subset B \rightarrow ||u||_B \leq C||u||_A$.

definition of $S[k]$ is complicated: we first construct norms $||.||_{S[k,\kappa]}$ associated with every integer $k$ and cap $\kappa \subset S^1$. To do so, we introduce *null-frame coordinates* $(t_\omega,\, x_\omega)$, $\omega \in S^1$, on space-time, whose definition is as follows:

$$t_\omega = \frac{1}{\sqrt{2}}(1,\omega)\cdot(t,x)$$

$$x_\omega = (t,x) - \frac{t_\omega}{\sqrt{2}}(1,\omega)$$

Thus these are Cartesian coordinates with respect to a tilted reference frame, whose 'time axis' with direction $\frac{1}{\sqrt{2}}(1,\omega)$ lies along the light cone. Now we introduce the space $PW[\kappa]$ defined as the atomic Banach space whose atoms are Schwartz functions $\psi \in \mathcal{S}(\mathbf{R}^{2+1})$ satisfying

$$\inf_{\omega\in\tilde{\kappa}} ||\psi||_{L^2_{t_\omega}L^\infty_{x_\omega}} \leq 1,$$

where $\tilde{\kappa}$ is a slightly grown version of $\kappa$ (say by a factor $\frac{11}{10}$) concentric with it. Thus for $\psi \in \mathcal{S}(\mathbf{R}^{2+1})$,we have

$$||\psi||_{PW[\kappa]} := \inf_{\int_{\tilde{\kappa}}\psi_\omega d\omega=\psi} \int_{\tilde{\kappa}} ||\psi_\omega||_{L^2_{t_\omega}L^\infty_{x_\omega}}\, d\omega$$

Moreover, we put for $\psi$ as above

$$||\psi||_{NFA[\kappa]^*} = \sup_{\omega\notin 2\kappa} \mathrm{dist}(\omega,\kappa)||\psi||_{L^\infty_{t_\omega}L^2_{x_\omega}}$$

Now we put

$$||\psi||_{S[k,\kappa]} = ||\psi||_{L^\infty_t L^2_x} + 2^{-\frac{k}{2}}|\kappa|^{-\frac{1}{2}}||\psi||_{PW[\kappa]} + |\psi||_{NFA[\kappa]^*}$$

This definition immediately entails the following fundamental first bilinear inequality

$$||\phi\psi||_{L^2_t L^2_x} \leq C\frac{|\kappa'|^{\frac{1}{2}}2^{\frac{k'}{2}}}{\mathrm{dist}(\kappa,\kappa')}||\phi||_{S[k,\kappa]}||\psi||_{S[k',\kappa']}, \tag{2.1}$$

provided $2\kappa \cap 2\kappa' = \emptyset$. We now construct the norms $||\psi||_{S[k]}$ by evaluating suitably microlocalized pieces of $\psi$ with respect to the $||.||_{S[k,\kappa]}$, taking a suitable mean and combining this with $||.||_{\dot{X}^{a,b,c}_k}$ type norms. The null-frame norms may be thought of as controlling the 'free wave-like' character of $\psi$, while the remaining norms may be thought of as controlling the 'elliptic character' of $\psi$.

For every integer $l < -10$, subdivide $S^1$ into a uniformly finitely overlapping collection $K_l$ of caps $\kappa$ of diameter $2^l$. Also, for every integer $\lambda$ with $-10 \geq \lambda \geq l$, we subdivide the angular sector $\{\xi \in \mathbf{R}^2 | \frac{\xi}{|\xi|} \in \kappa,\ |\xi| \sim 2^k\}$ into a uniformly finitely overlapping collection $C_{k,\kappa,\lambda}$ of slabs $R$ of width $2^{k+\lambda}$. We introduce various localization operators associated with these regions: for each $\kappa \in K_l$, choose a smooth cutoff $a_\kappa : S^1 \to \mathbf{R}_{\geq 0}$ supported on a dilate of $\kappa$. These are to be chosen such that $\sum_{\kappa\in K_l} a_\kappa = 1$. We also introduce cutoffs $m_R(.) : \mathbf{R}_{>0} \to \mathbf{R}_{\geq 0}$ such that the cutoff $m_R(|\xi|)a_\kappa(\frac{\xi}{|\xi|})$ localizes to a dilate of the slab $R$. Also, we require that $\sum_{R\in C_{k,\kappa,\lambda}} m_R(|\xi|) = \chi_0(\frac{|\xi|}{2^k})$. We have the associated pseudo differential operator $\tilde{P}_R\psi$:

$$\widehat{\tilde{P}_R\psi}(t,\,\xi) = m_R(|\xi|)a_\kappa(\frac{\xi}{|\xi|})\hat{\psi}(\xi)$$

We also have the $\Psi$DO's $P_{k,\kappa}$ associated with multiplier $a_\kappa(\frac{\xi}{|\xi|})\chi_0(\frac{|\xi|}{2^k})$. Then, almost[2] as in [**23**] we define

$$\begin{aligned}||\psi||_{S[k]} :=&||\psi||_{L_t^\infty L^2} + ||\psi||_{\dot{X}_k^{0,\frac{1}{2},\infty}} + ||\psi||_{\dot{X}_k^{-\frac{1}{2},1,2}} \\ &+\sup_{\pm}\sup_{l<-10}\sup_{-10\geq\lambda\geq l}|\lambda|^{-1}(\sum_{\kappa\in K_l}\sum_{R\in C_{k,\kappa,\lambda}}||\tilde{P}_R Q^{\pm}_{<k+2l}\psi||^2_{S[k,\pm\kappa]})^{\frac{1}{2}}.\end{aligned} \tag{2.2}$$

This norm looks very complicated, but it isn't too hard to get control over its ingredients. A fundamental inequality [**23**] for example states that

$$||P_k Q_{<k+O(1)}\psi||_{S[k]} \leq C||P_k\psi||_{\dot{X}_k^{0,\frac{1}{2},1}} \tag{2.3}$$

We also need a norm $||.||_N$ with respect to which we evaluate the nonlinearities of our wave equations . We shall put

$$||\psi||_N := \sqrt{\sum_{k\in\mathbf{Z}}||P_k\psi||^2_{N[k]}},$$

where the $N[k]$ will be constructed as atomic Banach spaces. More precisely, we let $N[k]$ be the atomic Banach space whose atoms are Schwartz functions $F\in\mathcal{S}(\mathbf{R}^{2+1})$ with spatial Fourier support contained in the region $|\xi|\sim 2^k$ and

(1) $||F||_{L_t^1\dot{H}^{-1}}\leq 1$ and $F$ has modulation $< 2^{k+100}$.

(2) $F$ is at modulation $\sim 2^j$ and satisfies $||F||_{L_t^2L_x^2}\leq 2^{\frac{j}{2}}2^k$.

(3) $F$ satisfies $||F||_{\dot{X}_k^{-\frac{1}{2},-1,2}}\leq 1$.

(4) There exists an integer $l<-10$, and Schwartz functions $F_\kappa$ with Fourier support in the region

$$\{(\tau,\xi)|\pm\tau>0,\,|\,|\tau|-|\xi|\,|\leq 2^{k-2l-100},\,\frac{\xi}{|\xi|}\in\pm\kappa\}$$

with the properties

$$F=\sum_{\kappa\in K_l}F_\kappa,\;(\sum_{\kappa\in K_l}||F_\kappa||^2_{NFA[\kappa]})^{\frac{1}{2}}\leq 2^k$$

In the last inequality, $NFA[\kappa]$ denotes the dual of $NFA[\kappa]^*$ (the completion of $\mathcal{S}(\mathbf{R}^{2+1})$ with respect to $||.||_{NFA[\kappa]^*}$) used in the definition of $S[k,\kappa]$: Thus $NFA[\kappa]$ is the atomic Banach space whose atoms $F$ satisfy

$$\frac{1}{\text{dist}(\omega,\,\kappa)}||F||_{L^1_{t_\omega}L^2_{x_\omega}}\leq 1$$

for some $\omega\notin 2\kappa$.

[2]The original definition also contained a norm $||P_kQ_{\geq k}\partial_t\psi||_{L_t^M\dot{H}^{-1+\frac{1}{M}}}$. Inspection of the proof there yields that this is superfluous, though, since the norm $||.||_{\dot{X}_k^{-\frac{1}{2},-1,2}}$ suffices for the elliptic estimates.

Observe that $\dot{X}_k^{-1,-\frac{1}{2},1} \subset N[k]$, $P_k(L_t^1 \dot{H}^{-1}) \subset N[k]$.

This definition immediately entails the fundamental 2nd bilinear inequality (again $2\kappa \cap 2\kappa' = \emptyset$)

$$||\phi\psi||_{NFA[\kappa]} \leq C\frac{2^{\frac{k'}{2}}|\kappa'|^{\frac{1}{2}}}{\text{dist}(\kappa,\kappa')}||\phi||_{L_t^2 L_x^2}||\psi||_{S[k',\kappa']} \tag{2.4}$$

We quickly summarize here the main properties of these spaces we shall need, all proved in [**23**]: all functions $\phi$, $\psi$ etc. below shall be in $\mathcal{S}(\mathbf{R}^{2+1})$.

(a): *Product type estimate*

$$||P_k Q_j(P_{k_1}\phi_1 P_{k_2}\phi_2)||_{\dot{X}_k^{0,\frac{1}{2},\infty}} \leq C 2^{\min\{k_1,k_2\}} 2^{\min\{\frac{j-\min\{k,k_1,k_2\}}{4+},0\}}$$
$$2^{\min\{\frac{\max\{k_1,k_2\}-j}{2},0\}}||P_{k_1}\phi_1||_{S[k_1]}||P_{k_2}\phi_2||_{S[k_2]}$$

(b): *Bilinear null-form estimate*: let $R_\nu$ denote the Riesz type operator $R_\nu = \frac{\partial_\nu}{\sqrt{-\triangle}}$. For $0 \leq p < \frac{1}{4}$, we have[3]

$$\begin{aligned}&||P_k[R_1 P_{k_1}\psi_1 R_2 P_{k_2}\psi_2 - R_2 P_{k_1}\psi_1 R_2 P_{k_2}\psi_2]||_{\dot{X}_0^{p,-p,2}}\\ &\qquad \leq C_p 2^{\frac{\min\{k_1,k_2,k\}}{2}}\prod_{i=1,2}||P_{k_i}\psi_i||_{S[k_i]}\end{aligned} \tag{2.5}$$

$$\begin{aligned}&||P_k[R_1 P_{k_1}\psi_1 R_\nu P_{k_2}\psi_2 - R_2 P_{k_1}\psi_1 R_\nu P_{k_2}\psi_2]||_{\dot{X}_0^{p,-p,2}}\\ &\qquad \leq C_p|k-k_1|^2 2^{\frac{\min\{k_1,k_2,k\}}{2}}\prod_{i=1,2}||P_{k_i}\psi_i||_{S[k_i]}\end{aligned}$$

(c): *Trilinear null-form estimates*: these arise upon formulating the derivative wave equations in the Coulomb Gauge and applying Hodge type decompositions, as explained below: let $I = \sum_{k\in\mathbf{Z}} P_k Q_{<k+100}$:

$$\begin{aligned}&||\partial^\beta P_0[R_\alpha P_{k_1}\psi_1 \triangle^{-1}\sum_{j=1}^{2}\partial_j I[R_\beta P_{k_2}\psi_2 R_j P_{k_3}\psi_3 - R_\beta P_{k_3}\psi_3 R_j P_{k_2}\psi_2]]\\ &+\partial_\alpha P_0[R_\beta P_{k_1}\psi_1 \triangle^{-1}\sum_{j=1}^{2}\partial_j I[R^\beta P_{k_2}\psi_2 R_j P_{k_3}\psi_3 - R_j P_{k_2}\psi_2 R^\beta P_{k_3}\psi_3]]||_{N[0]}\\ &\leq C 2^{\delta_1\min\{-\min\{k_1,k_2,k_3\},0\}}\prod_i 2^{\delta_2\min\{\max_{j\neq i}\{k_i,k_i-k_j\},0\}}\prod_l ||P_{k_l}\psi_l||_{S[k_l]},\end{aligned} \tag{2.6}$$

$$\begin{aligned}&||P_0\partial^\beta[R_\beta P_{k_1}\psi_1 \triangle^{-1}\sum_j \partial_j I[R_\alpha P_{k_2}\psi_2 R_j P_{k_3}\psi_3 - R_j P_{k_2}\psi_2 R_\alpha P_{k_3}\psi_3]]||_{N[0]}\\ &\leq C 2^{\delta_1\min\{-\min\{k_1,k_2,k_3\},0\}}\prod_i 2^{\delta_2\min\{\max_{j\neq i}\{k_i,k_i-k_j\},0\}}\prod_l ||P_{k_l}\psi_l||_{S[k_l]}.\end{aligned} \tag{2.7}$$

Of course one may rescale these, i. e. replace $P_0$ by $P_k$, $k \in \mathbf{Z}$. Then one needs to replace $\min\{-\min\{k_1,k_2,k_3\},0\}$ by $\min\{-\min\{k_1-k,k_2-k,k_3-k\},0\}$ etc.

---

[3] Moreover, applying an operator $Q_l$ in front, where $l >> k$, we can include factors $2^{\frac{\min\{l-k_2,0\}}{2}}$ on the right hand side.

(d): *'energy inequality'* The following relates the spaces $S[k]$ and $N[k]$:

$$||P_k\psi||_{S[k]([-T,T]\times\mathbf{R}^2)} \leq C \inf_{0<T_0\leq T}[\min\{2^k T_0, 1\}^{-\frac{1}{2}}||\Box P_k\psi||_{N[k]([-T,T]\times\mathbf{R}^2)} + \sup_{t_0\in[-T_0,T_0]}||P_k\psi[t_0]||_{L^2\times\dot{H}^{-1}}]. \tag{2.8}$$

In this inequality, one can leave out the factor $\min\{2^k T_0, 1\}^{-\frac{1}{2}}$ and replace $S[k]$ with the following stronger norm

$$||\psi||_{S'[k]} := 2^{-k}||\nabla_{x,t}\psi||_{L_t^\infty L_x^2} + ||\nabla_{x,t}\psi||_{\dot{X}_k^{-1,\frac{1}{2},\infty}} + \sup_{\pm}\sup_{l<-10}\sup_{-10\geq\lambda\geq l}|\lambda|^{-1}(\sum_{\kappa\in K_l}\sum_{R\in C_{k,\kappa,\lambda}}||\tilde{P}_R Q^{\pm}_{<k+2l}\psi||^2_{S[k,\pm\kappa]})^{\frac{1}{2}},$$

provided the norm $N[k]$ is replaced by $L_t^1\dot{H}^{-1}$ or $\dot{X}_k^{-1,-\frac{1}{2},1}$.

(e): *Relation to Strichartz type spaces:* Let $p$, $q$ satisfy $\frac{1}{p}+\frac{1}{2q}<\frac{1}{4}$. Then we have

$$||P_0\psi||_{L_t^p L_x^q} \leq C_{p,q}||P_0\psi||_{S[0]}$$

(f): *Relation to improved Strichartz type estimates:* for $l < -10$, let $C_{0,l}$ be a covering of the frequency region $|\xi| \sim 1$ by uniformly finitely overlapping discs of radius $\sim 2^l$. Let $P_{0,c}$ localize the Fourier support to the disc $c$, such that $\sum_{c\in C_{0,l}} P_c = P_0$. Let $8 \geq p > 4$. Then we have

$$(\sum_{c\in C_{0,l}}||P_{0,c}\psi||^2_{L_t^p L_x^q})^{\frac{1}{2}} \leq C_p 2^{(\frac{3}{4+}-\frac{2}{p})l}||\psi||_{S[0]}$$

(g): *Bilinear inequality relating the $S[k]$, $N[k]$:* Let $F$, $\psi \in \mathcal{S}(\mathbf{R}^{2+1})$. Then for $j \leq \min\{k_{1,2}\}+O(1)$ we have

$$||P_k[P_{k_1}\psi P_{k_2}Q_j F]||_{N[k]} \leq C2^{\delta(j-\min\{k_1,k_2\})}2^{\min\{k_1-k_2,0\}}||P_{k_1}\psi_1||_{S[k_1]}||P_{k_2}F||_{\dot{X}_{k_2}^{0,-\frac{1}{2},\infty}}$$

If $k_1 = k_2 + O(1)$ $j \leq r+k$ for some $r \leq 0$, we have

$$||P_k Q_{<r+k}[P_{k_1}Q_j F P_{k_2}Q_{<2k+r-k_1}\psi]||_{N[k]} \leq C2^{\delta r}||F||_{\dot{X}_{k_1}^{0,-\frac{1}{2},\infty}}||P_{k_2}\psi_2||_{S[k_2]}.$$

**2.0.4. A modification of the spaces; Moser type estimates.** In spite of the above properties, the spaces $S[k]$ don't appear flexible enough to handle Moser type estimates of the kind we shall need. More precisely, the property

$$||\psi_1 A(\nabla^{-1}\psi_2)||_S \leq CC(||\psi_1||_S, ||\psi_2||_S)$$

where $\psi_{1,2} \in \mathcal{S}(\mathbf{R}^2)$, $A(.) \in C^\infty(\mathbf{R})$ with bounded derivatives, and $\nabla^{-1}$ schematic notation for linear combinations of operators of the form $\triangle^{-1}\partial_j$, appears violated. This is a consequence of the fact that the product estimate 3.4(a) does not allow us to recover enough exponential gains in the difference $k - k_1$ if $k_1 >> k$, the high-high interaction case.

One way around this would be to re-engineer the way functions get subdivided into 'free wave parts' and 'elliptic parts'. Indeed, one has better product estimates than 3.4(a) for free waves, see Klainerman-Foschi[4] **[11]**. Our way here around this shall

[4]It appears, however, that even for free waves, a high-high-low interaction resulting in an elliptic product (Fourier support very far from the light cone) does not lead to the desired exponential gain in the frequencies.

be to 'enlarge the space $S[k]$', shrinking the norm $||.||_{S[k]}$ suitably. More precisely, we analyze the 'bad high-high' frequency interactions and observe that by virtue of the spherical symmetry assumption, these cases are actually favorable in some sense. Indeed, we shall be able to exploit the well-known fact (e. g. **[30]**, **[32]**) that the range of admissible Strichartz estimates is significantly improved in this situation:

THEOREM 2.1. *Let $\psi \in \mathcal{S}(\mathbf{R}^{2+1})$ be invariant under rotations. Then we have the inequality*

$$||P_0\psi||_{L_t^pL_x^q} \leq C||\psi[0]||_{L_t^\infty L_x^2} + ||\Box\psi||_{L_t^1L_x^2}$$

*provided the condition $\frac{1}{p}+\frac{1}{q}<\frac{1}{2}$ holds.*

We note that theorem 2.1 implies easily the following corollary:

COROLLARY 2.1. *The derivative components $\psi_\nu$ satisfy the estimates*

$$||P_k\psi_\nu||_{L_t^pL_x^q} \leq Cc_k$$

*for a system of numbers $\{c_k\}_{k\in\mathbf{Z}}$ ('frequency envelope') which satisfies $\sum_{k\in\mathbf{Z}} 2^{\delta|k|}c_k < \infty$, provided the condition $\frac{1}{p}+\frac{1}{q}<\frac{1}{2}$ holds and $\delta>0$ is sufficiently small.*

PROOF. This follows by applying a simple frequency trichotomy to the frequency localized expression for $\psi_\nu$ in terms of $\frac{\partial_\nu \mathbf{x}}{\mathbf{y}}$, $\frac{\partial_\nu \mathbf{y}}{\mathbf{y}}$. The latter are controlled by application of theorem 2.1 as well as Corollary 1.1. □

DEFINITION 2.2. : We put

$$||\psi||_L := \sup_{(p,q)|\frac{1}{p}+\frac{1}{2q}\leq\frac{1}{4}} 2^{k(\frac{1}{p}+\frac{2}{q}-1)} \sup_{l\leq 0} 2^{-\max\{\frac{1}{2+}-\frac{1}{q},0\}l}\Big(\sum_{c\in C_{k,l}} ||P_c\psi||^2_{L_t^pL_x^q}\Big)^{\frac{1}{2}}$$

Here $C_{k,l}$ is a finitely overlapping cover of the frequency region $|\xi|\sim 2^k$ by discs of radius $2^{k+l}$, with associated Fourier localizers $P_c$, $c\in C_{k,l}$. Also, put $2+ = 2+\frac{1}{1000}$, say, and let $\mu$ be a small positive number, say $\frac{1}{1000}$. We let $\mathcal{S}[k]$ be the atomic Banach space whose atoms satisfy one of the following:

(1) *Type 1 atoms*: Fix $\delta_0$ small, say $\delta_0 = \frac{1}{1000}$. These are functions $\psi \in \mathcal{S}(\mathbf{R}^{2+1})$ satisfying

$$||\psi||_{A[k]} = ||\psi||_L + ||\psi||_{\dot{X}_k^{0,\frac{1}{2},\infty}} + ||\psi||_{\dot{X}_k^{-\frac{1}{2},1,2}} + \sup_{\frac{1}{p}+\frac{1}{q}<\frac{1}{2}-\delta_0} 2^{(\frac{1}{p}+\frac{2}{q}-1)k}||\psi||_{L_t^pL_x^q}$$

$$+\sup_{\pm}\sup_{l<-10}\sup_{-10\geq\lambda\geq l}|\lambda|^{-1}\Big(\sum_{\kappa\in K_l}\sum_{R\in C_{k,\kappa,\lambda}}||\tilde{P}_RQ^{\pm}_{<k+2l}\psi||^2_{S[k,\pm\kappa]}\Big)^{\frac{1}{2}} \leq 1$$

(2) *Atoms of the 2nd type*: Let $I=\sum_{k\in\mathbf{Z}}P_kQ_{<k+100}$. Then $\psi\in\mathcal{S}(\mathbf{R}^{2+1})$ is of the 2nd type provided

$$||\psi||_{B[k]} := \sup_{\frac{1}{p}+\frac{1}{q}<1-10\delta_0} 2^{k[\frac{1}{p}+\frac{2}{q}-1]}||\psi||_{L_t^pL_x^q} + ||\psi||_L$$

$$+ ||(1-I)\psi||_{\dot{X}_k^{-(\frac{1}{2}-\mu),1-\mu,1}} \leq 1,$$

The range of Lebesgue exponents $(p,q)$ includes the pairs$(1+,\infty)$, $(\infty,1+)$.

Note that any function $\psi \in \mathcal{S}(\mathbf{R}^{2+1})$ may be decomposed into two pieces $\psi = \alpha + \beta$ satisfying

$$||\alpha||_{A[k]} \leq C||\psi||_{S[k]},\ ||\beta||_{B[k]} \leq C||\psi||_{S[k]}$$

We call $\alpha$ 'of first type' and $\beta$ 'of 2nd type'. We let

$$||\psi||_S := (\sum_{k\in\mathbf{Z}} ||P_k\psi||^2_{S[k]})$$

Unfortunately, these norms do not quite suffice to close all the estimates. The following theorem contains some bilinear estimates, which we were unable to build into a linear framework. These have to be proved independently. To state the theorem, we use the concept of *frequency envelope:* following Tao [**36**], we call a sequence of positive numbers $\{c_k\}_{k\in\mathbf{Z}}$ a frequency envelope provided $C^{-1}c_b 2^{-\sigma|a-b|} \leq c_a \leq C2^{\sigma|a-b|}c_b$ for some $\sigma > 0$ and $C >> 1$.

THEOREM 2.3. *Let $\psi_{1,2} \in \mathcal{S}(\mathbf{R}^{2+1})$. Assume that $||P_k\psi_{1,2}||_{S[k]} \leq Cc_k$ for a sufficiently flat frequency envelope $\{c_l\}$. Let*

$$||\nabla^{-1}\psi_2||_{L_t^\infty L_x^\infty} \leq C_0$$

*More precisely, assume that for each $k \in \mathbf{Z}$ one may split $P_k\psi_2 = \alpha_k + \beta_k$ into functions of first and 2nd type, respectively, such that $||\nabla^{-1}\sum_k \alpha_k||_{L^\infty} \leq C_0$, $||\nabla^{-1}\sum_k \beta_k||_{L^\infty} \leq C_0$. Also assume that $\psi_2$ satisfies the bilinear estimates stated further below*[5]*. Then we can conclude*[6] *$\forall k \in \mathbf{Z}$*

$$||P_k[\psi_1\nabla^{-1}\psi_2]||_{S[k]} \leq Cc_k$$

*In particular, if $A(.) : \mathbf{R} \to \mathbf{C}$ is real analytic with bounded derivatives of arbitrary order, and $\psi_2$ real valued, we can conclude that $\forall k \in \mathbf{Z}$*

$$||P_k[\psi_1 A(\nabla^{-1}\psi_2)]||_{S[k]} \leq Cc_k$$

*A more precise version is as follows: for suitable $\delta > 0$,*

$$||P_k[P_{k_1}\psi_1\nabla^{-1}P_{k_2}\psi_2]||_{S[k]} \leq C2^{-\delta|k-k_1|}||P_{k_1}\psi_1||_{S[k_1]}||P_{k_2}\psi_2||_{S[k_2]}$$

*Now assume that $||\psi_{1,2}||_S + ||\psi_{1,2}||_{\dot{B}_2^{0,1}} < C_0$. Then we can conclude*

$$||P_k[\psi_1 A(\nabla^{-1}\psi_2)]||_{\dot{X}_k^{0,\frac{1}{2},\infty}} \leq C,$$

*for a suitable $C = C(C_0)$. More precisely, decomposing*

$$P_k[\psi_1 A(\nabla^{-1}\psi_2)] = \alpha + \beta$$

*into functions of first and 2nd type, respectively, we may assume that*

$$||R_0(\alpha+\beta)||_{L_t^\infty L_x^2} \leq C,\ ||\beta||_{\dot{X}_k^{0,\frac{1}{2},1}} \leq C$$

---

[5]Substitute $\psi_2$ instead of $\psi_1 A(\nabla^{-1}\psi_2)$.

[6]The implied constants in the statements below depend on $C_0$, the constant $C$ in the bilinear estimates below as well as the decay of the frequency envelope and the constants chosen in the definition of $\mathcal{S}[k]$.

*for* $C = C(C_0, \sup_{k\in\mathbf{Z}} ||R_0 P_k \psi_1||_{L_t^\infty L_x^2}, \sup_{k\in\mathbf{Z}} ||R_0 P_k \psi_2||_{L_t^\infty L_x^2})$. *Next, we have the bilinear estimates*

$$\sup_{\phi|\,||\phi||_{S[k_3]}\leq 1} \sup_{k_{1,2,3}} \sup_{l<-10} 2^{-\frac{\min\{k_1,k_3\}}{2}} 2^{-\delta_1 l}$$

$$\Big(\sum_{c\in C_{k_3,l}} ||P_{k_1}[P_{k_2} R_0[\psi_1 A(\nabla^{-1}\psi_2)] P_c\phi]||^2_{L_t^2 L_x^2}\Big)^{\frac{1}{2}} \leq CC$$

$$\sup_{\phi|\,||\phi||_{S[k_3]}\leq 1} \sup_{k_{1,2}\in\mathbf{Z}} 2^{\frac{k_2(1-3\mu)}{2+2\mu}} ||P_{k_1}[P_{k_2} R_0[\psi_1 A(\nabla^{-1}\psi_2)] P_{k_2+O(1)}\phi]||_{L_t^2 L_x^{1+\mu}} \leq C,$$

*where* $C = C(||\psi_1||_S, ||\psi_2||_S, ||\psi_2||_{\dot{B}_2^{0,1}})$*, recall the definition in 3.4. The same estimates hold provided one replaces* $A(\nabla^{-1}\psi_2)$ *by* $A(\nabla^{-1}(\psi_2\nabla^{-1}\psi_3))$*, where* $\psi_3$ *satisfies similar estimates as* $\psi_2$*.*

The proof of this is a long calculation deferred to an appendix.

**2.0.5. Proof of the Theorem 1.10.** To show existence, it suffices to show that some subcritical norm $\sum_{\nu=0}^{2} ||\delta\psi_\nu(t)||_{H^\delta}$ is globally bounded in $t$. Indeed, reasoning exactly as in [**23**], one deduces that for every finite time interval $[-T, T]$, one has with $\tilde{u} = (\tilde{\mathbf{x}}, \tilde{\mathbf{y}})$ and $\tilde{\delta} < \delta$:

$$\sup_{t\in[-T,T]} \sum_{\nu=0}^{2} [||\tilde{\mathbf{x}}(t)||_{H^{1+\tilde{\delta}}} + ||\tilde{\mathbf{y}}(t)||_{H^{1+\tilde{\delta}}} + ||\partial_t\tilde{\mathbf{x}}(t)||_{H^{\tilde{\delta}}} + ||\partial_t\tilde{\mathbf{y}}(t)||_{H^{\tilde{\delta}}}] < \infty$$

Using the subcritical result of
Klainerman-Machedon [**12**], one deduces from here that there can't be breakdown of smoothness after finite time.
Global boundedness of a subcritical norm in turn shall follow from the following Bootstrap Proposition: to formulate it, we shall need time-localized versions of the spaces $S[k]$: for $\psi \in C_0^\infty([-T, T] \times \mathbf{R}^2)$ define

$$||\psi||_{S[k]([-T,T]\times\mathbf{R}^2)} := \inf_{\tilde{\psi}\in\mathcal{S}(\mathbf{R}^{2+1})|\tilde{\psi}|_{[-T,T]\times\mathbf{R}^2}=\psi} ||\tilde{\psi}||_{S[k]}$$

We use similar definitions for $||.||_{S([-T,T]\times\mathbf{R}^2)}$ $||\psi||_{S([-T,T]\times\mathbf{R}^{2+1})}$ etc. and also different time intervals $[T_1, T_2]$ etc.

PROPOSITION 2.1. *In the situation of Theorem 1.10, let the smooth Wave Map extending* $\tilde{u}[0]$ *exist on the time interval* $[-T, T]$*. There exists* $T_1 > 0$ *such that for* $T \geq T_1$ *and every* $K > 0$ *sufficiently large, there exists* $\epsilon > 0$ *such that the following conclusion applies: Introduce the frequency envelope*

$$\tilde{c}_k := \sup_{t\in[-T_1,T_1]} \sum_{k_1\in\mathbf{Z}} 2^{-\sigma|k-k_1|} ||P_{k_1}\delta\psi_\nu(t)||_{L_x^2} + \epsilon c_k,$$

*where* $\{c_k\}$ *is as in the proof of lemma 2.6, and assume*
$\sup_{t\in[-T_1,T_1]} \sum_{\nu=0,1,2} ||\delta\psi_\nu(t)||_{L_x^2} \leq \epsilon$*. Then for any* $T \geq T_1$ *we have*

$$\sup_\nu ||P_k\delta\psi_\nu||_{S[k]([T_1,T]\times\mathbf{R}^2)} < K\tilde{c}_k \Rightarrow ||P_k\delta\psi_\nu||_{S[k]([T_1,T]\times\mathbf{R}^2)} < \frac{K}{2}\tilde{c}_k.$$

*A similar inequality holds by replacing* $T$, $T_1$ *by* $-T$, $-T_1$*.*

Assuming this for now, we continue with the proof of Theorem 1.10. We claim that by local well-posedness of (1.1), (1.2), there exists $\epsilon = \epsilon(\mu) > 0$ such that (using terminology of Theorem 1.10) $||u[0] - \tilde{u}[0]||_{H^{1+\mu}\times H^{\mu}} < \epsilon$ implies that $\tilde{u}$ extends smoothly to $[-T_1, T_1]$, where $T_1$ is as in the preceding Proposition. To see that this is possible, we shall apply an inequality of Klainerman-Selberg to the equation satisfied by the differences $\tilde{\mathbf{x}} - \mathbf{x}$, $\tilde{\mathbf{y}} - \mathbf{y}$ of the coordinate representations of the perturbed and the spherically symmetric Wave Map, $\tilde{u} = (\tilde{\mathbf{x}}, \tilde{\mathbf{y}})$ and $u = (\mathbf{x}, \mathbf{y})$. Subdivide the interval $[-T_1, T_1]$ into small subintervals $I_i$, for which[7] $||(\mathbf{x}, \mathbf{y})||_{\mathcal{X}^{1+\mu,\theta}(I_i\times\mathbf{R}^2)} \leq C1$. This is possible by Corollary 1.1 and local well-posedness of (1.1), (1.2) in $H^{1+\mu}$. Note that

$$\Box(\ln\tilde{\mathbf{y}} - \ln\mathbf{y}) = (\frac{\partial_\nu\tilde{\mathbf{x}}}{\tilde{\mathbf{y}}} - \frac{\partial_\nu\mathbf{x}}{\mathbf{y}})\frac{\partial^\nu\tilde{\mathbf{x}}}{\tilde{\mathbf{y}}} + \frac{\partial^\nu\mathbf{x}}{\mathbf{y}}(\frac{\partial_\nu\tilde{\mathbf{x}}}{\tilde{\mathbf{y}}} - \frac{\partial_\nu\mathbf{x}}{\mathbf{y}})$$

with a similar equation holding for $\Box(\frac{\tilde{\mathbf{x}}}{\tilde{\mathbf{y}}} - \frac{\mathbf{x}}{\mathbf{y}})$. We deduce that upon denoting $I_i = [a_i, a_{i+1}]$, we have

$$\sum_\nu ||\frac{\partial_\nu\tilde{\mathbf{y}}}{\tilde{\mathbf{y}}} - \frac{\partial_\nu\mathbf{y}}{\mathbf{y}}||_{\mathcal{X}^{\mu,\theta}(I_i\times\mathbf{R}^2)} + ||\partial_\nu(\frac{\tilde{\mathbf{x}}}{\tilde{\mathbf{y}}}) - \partial_\nu(\frac{\tilde{\mathbf{x}}}{\tilde{\mathbf{y}}})||_{\mathcal{X}^{\mu,\theta}(I_i\times\mathbf{R}^2)}$$

$$\leq C\sum_\nu ||\frac{\partial_\nu\tilde{\mathbf{y}}}{\tilde{\mathbf{y}}} - \frac{\partial_\nu\mathbf{y}}{\mathbf{y}}||_{H^\mu(a_i)} + ||\partial_\nu(\frac{\tilde{\mathbf{x}}}{\tilde{\mathbf{y}}}) - \partial_\nu(\frac{\tilde{\mathbf{x}}}{\tilde{\mathbf{y}}})||_{H^\mu(a_i)}$$

$$+ |I_i|^{\epsilon(\mu)}[||\ln\tilde{\mathbf{y}} - \ln\mathbf{y}||_{\mathcal{X}^{1+\mu,\theta}(I_i\times\mathbf{R}^2)} + ||\frac{\tilde{\mathbf{x}}}{\tilde{\mathbf{y}}} - \frac{\mathbf{x}}{\mathbf{y}}||_{\mathcal{X}^{1+\mu,\theta}(I_i\times\mathbf{R}^2)}]$$

$$+ [|I_i|^{\epsilon(\mu)}||\ln\tilde{\mathbf{y}} - \ln\mathbf{y}||^3_{\mathcal{X}^{1+\mu,\theta}(I_i\times\mathbf{R}^2)} + ||\frac{\tilde{\mathbf{x}}}{\tilde{\mathbf{y}}} - \frac{\mathbf{x}}{\mathbf{y}}||^3_{\mathcal{X}^{1+\mu,\theta}(I_i\times\mathbf{R}^2)}]$$

We have used here the fact, due to Klainerman-Selberg, that

$$||\phi||_{\mathcal{X}^{s,\theta}([-T,T]\times\mathbf{R}^2)} \leq C||\phi[0]||_{H^s\times H^{s-1}} + T^\epsilon||\Box\phi||_{X^{s-1,\theta-1}([-T,T]\times\mathbf{R}^2)},\ \theta > \frac{1}{2}, s > 1$$

as well as the following inequality of Klainerman-Machedon

$$||\partial_\nu u_1\partial^\nu u_2||_{X^{s-1,\theta-1}} \leq C||u_1||_{\mathcal{X}^{s,\theta}}||u_2||_{\mathcal{X}^{s,\theta}}$$

Refining the subdivision $[-T_1, T_1] = \bigcup_{i=1}^N I_i$, $N = N(u, \mu)$, if necessary, we see that

$$||\sum_\nu ||\frac{\partial_\nu\tilde{\mathbf{y}}}{\tilde{\mathbf{y}}} - \frac{\partial_\nu\mathbf{y}}{\mathbf{y}}||_{H^\mu(a_{i+1})} + ||\partial_\nu(\frac{\tilde{\mathbf{x}}}{\tilde{\mathbf{y}}}) - \partial_\nu(\frac{\tilde{\mathbf{x}}}{\tilde{\mathbf{y}}})||_{H^\mu(a_{i+1})}$$

$$\leq C2[||\sum_\nu ||\frac{\partial_\nu\tilde{\mathbf{y}}}{\tilde{\mathbf{y}}} - \frac{\partial_\nu\mathbf{y}}{\mathbf{y}}||_{H^\mu(a_i)} + ||\partial_\nu(\frac{\tilde{\mathbf{x}}}{\tilde{\mathbf{y}}}) - \partial_\nu(\frac{\tilde{\mathbf{x}}}{\tilde{\mathbf{y}}})||_{H^\mu(a_i)}],$$

provided the quantity on the right hand side is less than some constant $c$. Thus if we choose $\epsilon < \frac{c}{2^N}$, we see that the Wave Map $\tilde{u}$ satisfying $||\tilde{u}[0] - u[0]||_{H^{1+\mu}\times H^\mu} << \epsilon$ will exist and be smooth on the interval $[-T_1, T_1]$. It follows from the argument just given and a simple algebra type estimate that by possibly shrinking the size of $||u[0] - \tilde{u}[0]||_{H^{1+\mu}\times H^\mu}$ we can ensure that

$$||\delta\psi_\nu||_{L_t^\infty H^\lambda([-T_1,T_1]\times\mathbf{R}^2)} \leq \epsilon,\ 0 < \lambda < \mu.$$

Now assume that the perturbed Wave Map $\tilde{u}$ breaks down at some time $T > T_1$. We claim that $\sup_{T_1\leq t<T}\sup_{k\in\mathbf{Z}}\tilde{c}_k^{-1}||P_k\delta\psi_\nu||_{S[k]([T_1,t]\times\mathbf{R}^2)} < \infty$. Indeed, in the

[7]The $\theta > \frac{1}{2}$ is chosen in dependence of $\mu$, see Klainerman-Selberg [**17**].

opposite case, choosing $K$ large enough(and if necessary shrinking $\epsilon$), and using the continuity of the function $t \to \sup_k \tilde{c}_k^{-1}||P_k\delta\psi_\nu||_{S[k]([T_1,t]\times\mathbf{R}^2)}$ for $t \in [T_1, T)$, see e. g. [**23**], it follows that there exists $T'$ satisfying the properties

$$\sup_{k\in\mathbf{Z}} \tilde{c}_k^{-1}||P_k\delta\psi_\nu||_{S[k]([T_1,T']\times\mathbf{R}^2)} = K,\ T_1 < T' < T$$

This, however, contradicts Proposition 2.1. This then implies that $\sup_{t<T}||P_k\delta\psi_\nu||_{S[k]([-t,t]\times\mathbf{R}^2)} \leq C\tilde{c}_k$. But by definition of $\tilde{c}_k$, $||.||_{S[k]}$, this implies that some subcritical norm $||\delta\psi_\nu||_{L_t^\infty H^\mu([-T,T]\times\mathbf{R}^2)} < \infty$, $\mu > 0$, which in turn implies that some $||\tilde{u}||_{L_t^\infty H^{1+\mu'}([-T,T]\times\mathbf{R}^2)} < \infty$. This in turn contradicts breakdown by the result of Klainerman-Machedon [**12**]. Of course, the preceding argument entails the bound $||\delta\psi_\nu||_{L_t^\infty L_x^2} \leq C\epsilon$. Indeed, one obtains that some range of subcritical norms $||.||_{H^\delta}$ satisfy that estimate.

**2.0.6. Proof of Proposition 2.1.** We first recall the following theorem from [**23**]: let $N(\{\psi_\nu\})$ be the nonlinearity on the right hand side of (1.8). Then we have

THEOREM 2.4. [**23**] *Let* $\psi_\nu \in C_0^\infty([-T,T]\times\mathbf{R}^2)$ *solve the combined system* (1.6), (1.7). *Then provided*

$$\sup_\nu ||\psi_\nu||_{S([-T,T]\times\mathbf{R}^2)} < K\epsilon$$

*and* $\epsilon$ *is sufficiently small in relation to* $K$, *we have*

$$||N(\{\psi_\nu\})||_{N([-T,T]\times\mathbf{R}^2)} \leq CK^3\epsilon^3$$

*More precisely, there exists a number* $\delta_1 > 0$ *such that we have*

$$||P_kN(\{\psi_\nu\})||_{N[k]([-T,T]\times\mathbf{R}^2)} \leq C(\sup_\nu \sup_{k_1\in\mathbf{Z}} 2^{-\delta_1|k-k_1|}||P_{k_1}\psi_\nu||_{S[k_1]([-T,T]\times\mathbf{R}^2)})K^2\epsilon^2$$

The proof in [**23**] of this relied on introduction of null-form structure into the nonlinearities by means of Hodge type decompositions, as briefly outlined in subsection 3.1. Thus writing $\psi_\nu = R_\nu\psi + \chi_\nu$ and requiring $\sum_{i=1,2}\partial_i\chi_i = 0$ results in

$$\chi_\nu = i\sum_{i,j=1}^{2}\partial_i\Delta^{-1}(\psi_\nu\Delta^{-1}\partial_j(\psi_i^1\psi_j^2 - \psi_j^1\psi_i^2) - \psi_i\Delta^{-1}\partial_j(\psi_\nu^1\psi_j^2 - \psi_j^1\psi_\nu^2)). \tag{2.9}$$

$$\psi = -\sum_{i=1,2} R_i\psi_i \tag{2.10}$$

One now writes the nonlinearities $N(\{\psi_\nu\})$ as sums of various terms which are gotten by substituting either gradient components $R_\nu\psi$ or elliptic components $\chi_\nu$ in place of $\psi_\nu$, substituting[8] the Schwartz extensions $\rho_\nu$ for $\psi_\nu$ which satisfy

$$||\rho_\nu||_{S([-T,T]\times\mathbf{R}^2)} \leq CK\epsilon,$$

and further microlocalizing constituents of the expressions thus obtained. One thereby obtains trilinear null-forms of the types recorded in 3.4(c). Substituting elliptic components $\chi_\nu$ results in terms at least quintilinear in the variables $\psi_\nu$, which are more elementary to estimate, but still appear to require null-form structure, which is obtained upon reiterating the Hodge type decomposition. One keeps going like this until the error terms obtained can be estimated without using null-structures, based only on Strichartz type estimates. Summarizing, we have

[8]One reexpresses $\psi$, $\chi_\nu$ in terms of $\psi_\nu$ via (2.9), (2.10).

THEOREM 2.5. [23] *Under the hypotheses of Theorem 2.4, we can construct a function*

$$\tilde{N}(\{\rho_\nu\}) \in \mathcal{S}(\mathbf{R}^{2+1})$$

*which is expressible as a sum of terms trilinear, quadrilinear etc. up to degree 11 in the* $\rho_\nu$*, and satisfies*

$$\tilde{N}(\{\rho_\nu\})|_{[-T,T]} = N(\psi_\nu)$$

$$||P_k\tilde{N}(\{\rho_\nu\})||_{S[k]} \leq C(\sup_\nu \sup_{k_1\in\mathbf{Z}} 2^{-\delta|k-k_1|}||P_{k_1}\rho_\nu||_{S[k_1]}) \max\{||\rho_\nu||_S^2, ||\rho_\nu||_S^{10}\}$$

We shall apply this theorem to our situation. The complication that arises here has to do with the fact that the estimates for $\{\psi_\nu\}$, the derivative components of the spherically symmetric Wave Map, are not with respect to $||.||_S$, but rather $||.||_{\mathcal{S}}$, in view of theorem 2.3. We state here

LEMMA 2.6. *For any* $\sigma > 0$ *sufficiently small, there exists a frequency envelope* $\{c_l\}_{l\in\mathbf{Z}}$ *with exponent* $\sigma$ *and* $\sum_{l\in\mathbf{Z}} c_l^2 \leq C1$ *such that* $\forall T > 0$ *we have*

$$||P_k\psi_\nu||_{\mathcal{S}[k]([-T,T]\times\mathbf{R}^2)} \leq Cc_k$$

*We can also assume* $\sum_l 2^{\mu|l|}c_l \leq C1$ *for* $\mu > 0$ *sufficiently small. Moreover, choosing a Schwartz extension* $\widetilde{P_k\psi_\nu}$ *of* $P_k\psi_\nu|_{[-T,T]}$ *satisfying the above estimates, we may decompose each* $\widetilde{P_k\psi_\nu}$ *into functions of first and 2nd type,* $\widetilde{P_k\psi_\nu} = \alpha_\nu + \beta_\nu$*, such that the following properties hold:*

$$||\alpha_\nu||_{A[k]} \leq Cc_k,\ ||\beta_\nu||_{B[k]} \leq Cc_k,\ ||\beta_\nu||_{\dot{X}_k^{0,\frac{1}{2},1}} \leq C,\ ||R_0(\alpha_\nu+\beta_\nu)||_{L_t^\infty L_x^2} \leq C$$

*where* $C$ *depends on* $||\frac{\mathbf{x}}{\mathbf{y}}||_{\dot{B}_2^{1,1}} + ||\ln\mathbf{y}||_{\dot{B}_2^{1,1}}$*. Moreover, the bilinear inequalities enunciated in theorem 2.3 hold for* $\psi_\nu$ *in place of* $[\psi_1 A(\nabla^{-1}\psi_2)]$ *there.*

PROOF. We define

$$c_k := \sum_{k_1\in\mathbf{Z}} 2^{-\sigma|k-k_1|}[||P_{k_1}N(...)||_{L_t^1L_x^2(\mathbf{R}^{1+2})} + ||P_{k_1}N(...)||_{L_t^2\dot{H}^{-\frac{1}{2}}}]$$

$$+\sum_{k_1\in\mathbf{Z}} 2^{-\sigma|k-k_1|}||P_{k_1}\frac{\partial_\nu\mathbf{y}}{\mathbf{y}}(0)||_{L_x^2} + \sup_{\nu=0,1,2}\sum_{k_1\in\mathbf{Z}} 2^{-\sigma|k-k_1|}||P_{k_1}\partial_\nu(\frac{\mathbf{x}}{\mathbf{y}})(0)||_{L_x^2},$$

where $N(...) = N(\nabla\mathbf{x}, \nabla\mathbf{y}, \mathbf{x}, \mathbf{y})$ runs over the nonlinearities in (1.1), (1.2). That this is indeed a frequency envelope with the desired properties follows from Corollary 1.1 as well as lemma 1.9. We need to exercise some care to get good enough control over the elliptic portions of $\psi_\nu$. For this, truncate $N(\nabla\mathbf{x}, \nabla\mathbf{y}, \mathbf{x}, \mathbf{y})$ past some time $T_0 >> \max\{2^{-k}, T\}$, and (committing abuse of notation) decompose the nonlinearity

$$P_kN(\nabla\mathbf{x}, \nabla\mathbf{y}, \mathbf{x}, \mathbf{y}) = P_kQ_{<k}N(\nabla\mathbf{x}, \nabla\mathbf{y}, \mathbf{x}, \mathbf{y}) + P_kQ_{\geq k}N(\nabla\mathbf{x}, \nabla\mathbf{y}, \mathbf{x}, \mathbf{y}).$$

Then consider

$$\Box^{-1}P_kQ_{\geq k}N(\nabla\mathbf{x}, \nabla\mathbf{y}, \mathbf{x}, \mathbf{y}),$$

where the operator $\Box^{-1}$ is division by the symbol $(\tau^2 - |\xi|^2)$ on the space-time Fourier side. Clearly, from definition we have

$$||\Box^{-1}P_kQ_{\geq k}\nabla_{x,t}N(\nabla\mathbf{x}, \nabla\mathbf{y}, \mathbf{x}, \mathbf{y})||_{L_t^1L_x^2} \leq C2^{-k}c_k$$

Thus there exists a time $t_0 < T_0$ with the property

$$||\Box^{-1}P_kQ_{\geq k}\nabla_{x,t}N(\nabla\mathbf{x}, \nabla\mathbf{y}, \mathbf{x}, \mathbf{y})(t_0)||_{L_x^2} \leq Cc_k$$

We easily check that (for $C$ independent of $k$)

$$||\Box^{-1}R_0P_kQ_{\geq k}\nabla_{x,t}N(\nabla\mathbf{x},\,\nabla\mathbf{y},\,\mathbf{x},\,\mathbf{y})||_{L_t^\infty L_x^2}\leq C,$$

while also (using the wave equation)

$$||R_0\nabla_{x,t}P_k(\frac{\mathbf{x}}{\mathbf{y}})||_{L_t^\infty L_x^2}+||R_0\nabla_{x,t}P_k\ln\mathbf{y}||_{L_t^\infty L_x^2}\leq C$$

Now construct a free wave $a$ with the properties

$$a(t_0)=P_k(\frac{\mathbf{x}}{\mathbf{y}})(t_0)-\Box^{-1}P_kQ_{\geq k}N(\nabla\mathbf{x},\,\nabla\mathbf{y},\,\mathbf{x},\,\mathbf{y})(t_0)$$

$$\partial_ta(t_0)=P_k\partial_t(\frac{\mathbf{x}}{\mathbf{y}})(t_0)-\Box^{-1}\partial_tP_kQ_{\geq k}N(\nabla\mathbf{x},\,\nabla\mathbf{y},\,\mathbf{x},\,\mathbf{y})(t_0)$$

and similarly for $\ln\mathbf{y}$. It follows that the quantity $\frac{\mathbf{x}}{\mathbf{y}}-a-\Box^{-1}P_kQ_{\geq k}N(...)$ satisfies

$$\Box(\frac{\mathbf{x}}{\mathbf{y}}-a-\Box^{-1}P_kQ_{\geq k}N(...))=P_kQ_{<k}N(\nabla\mathbf{x},\,\nabla\mathbf{y},\,\mathbf{x},\,\mathbf{y}),$$

$$||R_0\nabla_{x,t}P_k(\frac{\mathbf{x}}{\mathbf{y}}-a-\Box^{-1}P_kQ_{\geq k}N(...))||_{L_t^\infty L_x^2}\leq C$$

as well as $||\nabla_{x,t}(\frac{\mathbf{x}}{\mathbf{y}}-a-\Box^{-1}P_kQ_{\geq k}N(...))(t_0)||_{L_x^2}\leq Cc_k$. One also verifies that

$$||\Box^{-1}\nabla_{x,t}P_kQ_{\geq k}N(\nabla\mathbf{x},\,\nabla\mathbf{y},\,\mathbf{x},\,\mathbf{y})||_{\dot{X}_k^{-\frac{1}{2},1,2}}\leq Cc_k,$$

which by Sobolev's inequality also implies control over $||.||_L$ as well as the Strichartz type norms $||.||_{L_t^pL_x^q}$, $\frac{1}{p}+\frac{1}{q}<\frac{1}{2}-\delta_0$, of this expression. Now one solves the wave equation for $\frac{\mathbf{x}}{\mathbf{y}}-a-\Box^{-1}P_kQ_{\geq k}N(...)$ with initial data given at time $t_0$. Using 3.4(d) (which in turn relies on a truncated Duhamel's formula, see **[23]**), one constructs Schwartz extensions $\widetilde{P_k(\frac{\mathbf{x}}{\mathbf{y}})}$, $\widetilde{P_k\ln\mathbf{y}}$ of $P_k(\frac{\mathbf{x}}{\mathbf{y}})|_{[-T,T]}$, $P_k\ln\mathbf{y}|_{[-T,T]}$, respectively, with the properties

$$||\nabla_{x,t}\widetilde{P_k(\frac{\mathbf{x}}{\mathbf{y}})}||_{A[k]}\leq Cc_k,\ ||\nabla_{x,t}\widetilde{P_k\ln\mathbf{y}}||_{A[k]}\leq Cc_k$$

as well as

$$||R_0\nabla_{x,t}[\widetilde{P_k(\frac{\mathbf{x}}{\mathbf{y}})}]||_{L_t^\infty L_x^2}\leq C,\ ||R_0\nabla_{x,t}\widetilde{P_k\ln\mathbf{y}}||_{L_t^\infty L_x^2}\leq C$$

where $C$ is independent of $k$. Using a partition of unity, one glues these extensions together to get Schwartz extensions $\widetilde{\frac{\mathbf{x}}{\mathbf{y}}}$, $\widetilde{\ln\mathbf{y}}$ of $\frac{\mathbf{x}}{\mathbf{y}}|_{[-T,T]}$, $\ln\mathbf{y}|_{[-T,T]}$ which satisfy

$$||P_k(\widetilde{\frac{\mathbf{x}}{\mathbf{y}}})||_{A[k]}\leq Cc_k$$

etc. Now one recalls that

$$\psi_\nu=(\frac{\partial_\nu\mathbf{x}}{\mathbf{y}}+i\frac{\partial_\nu\mathbf{y}}{\mathbf{y}})e^{i\sum_{j=1,2}\triangle^{-1}\partial_j(\frac{\partial_j\mathbf{x}}{\mathbf{y}})},$$

plugs in the Schwartz extensions of $\frac{\mathbf{x}}{\mathbf{y}}$ etc. and uses theorem 2.3 to obtain the desired conclusion. $\Box$

Continuing with the proof of Proposition 2.1, our strategy now will be to analyze the wave equation satisfied by $\delta\psi_\nu = \tilde{\psi}_\nu - \psi_\nu$. Using (1.8) for both $\psi_\nu$, $\tilde{\psi}_\nu$, and subtracting, we obtain a first version. We eliminate $\tilde{\psi}_\nu$ by substituting $\delta\psi_\nu + \psi_\nu$. One thereby obtains a sum of products of components $\delta\psi_\nu$, $\psi_\nu$ which are at least linear in $\delta\psi_\nu$. Proceeding as in the previous description, we decompose the $\delta\psi_\nu$, $\psi_\nu$ into gradient and elliptic parts. For the $\delta\psi_\nu$, this is obtained by applying the procedure to $\tilde{\psi}_\nu$, $\psi_\nu$ and forming the difference, resulting in

$$\delta\psi_\nu = R_\nu(\delta\psi) + \delta\chi_\nu,\ \delta\psi = -\sum_{i=1,2} R_i\psi_i$$

$$\delta\chi_\nu = i\sum_{i,j=1}^{2} \partial_i\Delta^{-1}(\tilde{\psi}_\nu\Delta^{-1}\partial_j(\tilde{\psi}_i^1\tilde{\psi}_j^2 - \tilde{\psi}_j^1\tilde{\psi}_i^2) - \tilde{\psi}_i\Delta^{-1}\partial_j(\tilde{\psi}_\nu^1\tilde{\psi}_j^2 - \tilde{\psi}_j^1\tilde{\psi}_\nu^2))$$
$$- i\sum_{i,j=1}^{2} \partial_i\Delta^{-1}(\psi_\nu\Delta^{-1}\partial_j(\psi_i^1\psi_j^2 - \psi_j^1\psi_i^2) - \psi_i\Delta^{-1}\partial_j(\psi_\nu^1\psi_j^2 - \psi_j^1\psi_\nu^2)).$$

Clearly one can reexpress the latter difference as a sum of terms linear, quadratic and cubic in the $\delta\psi_\nu$, eliminating the $\tilde{\psi}_\nu = \delta\psi_\nu + \psi_\nu$. In order to demonstrate Proposition 2.1, we shall rely on the following refined

PROPOSITION 2.2. *Let*

$$\Box\delta\psi_\alpha = N_\alpha(\delta\psi_\nu,\ \psi_\nu)$$

*on $[-T,T]$. Proceeding as above, express the nonlinearity as a sum of trilinear null-forms (substituting the gradient components for $\delta\psi_\nu$, $\psi_\nu$), as well as error terms 'at least quintilinear' in $\delta\psi_\nu$, $\psi_\nu$ (which arise upon substituting $\delta\chi_\nu$, $\chi_\nu$). Denote the sum of terms which are linear in $\delta\psi_\nu$ by $N_{1\alpha}(\delta\psi_\nu,\ \psi_\nu)$. Then for any $\zeta > 0$ there exists $T_0 > 0$, such that for any fixed smooth function $\chi(t) \in C^\infty(\mathbf{R})$ with $supp(\chi) \subset [-1,1]^c$, $\chi|_{[-2,2]^c} = 1$, we have*

$$||P_k(\delta\psi_\nu)||_{S[k]([-T,T]\times\mathbf{R}^2)} \leq K\tilde{c}_k,\ T > \tilde{T}_0 \geq T_0$$
$$\Rightarrow ||\chi(\frac{t}{\tilde{T}_0})P_kN_{1\alpha}(\delta\psi_\nu,\ \psi_\nu)||_{N[k]([-T,T]\times\mathbf{R}^2)} \leq C\zeta K\tilde{c}_k$$

*Here $\{\tilde{c}_k\}$ is associated with $\tilde{T}_0$ as in Proposition 2.1 (substitute $\tilde{T}_0$ for $T_1$). Moreover, denoting the terms at least quadratic in $\delta\psi_\nu$ by $N_2(\delta\psi_\nu,\psi_\nu)$, and letting $\epsilon, \tilde{c}_k$ be as in the statement of Proposition 2.1, the following conclusion holds provided $\epsilon$ is small enough and $\{\tilde{c}_k\}$ 'flat enough':*

$$||P_k(\delta\psi_\nu)||_{S[k]([-T,T]\times\mathbf{R}^2)} \leq K\tilde{c}_k \Rightarrow ||P_kN_{2\alpha}(\delta\psi_\nu,\psi_\nu)||_{N[k]([-T,T]\times\mathbf{R}^{2+1})} \leq C\epsilon K^2c_k.$$

Deferring the proof of this for the moment, we continue with the proof of Proposition 2.1. Let $\zeta < 1/C$ for some $C >> 1$, and construct $T_0$ as in Proposition 2.2; Define $T_1 := 2T_0$. Now assume we have the situation in the statement of Proposition 2.1. We intend to use the energy inequality 3.4(d). Fix $k \in \mathbf{Z}$, and consider $P_k\delta\psi_\nu$. We distinguish between the cases $T - T_1 < \frac{2^{-k}}{C}$ and the opposite. In the former case, the wave equation becomes useless, and we use the divergence-curl system directly: observe that by virtue of (1.6) we have for $i = 1, 2$, $T_1 < t \leq T$

$$P_k\delta\psi_i(t) - P_k\delta\psi_i(T_1) = \int_{T_1}^{t} \partial_iP_k\delta\psi_i dt + \int_{T_1}^{t} P_k[N(\psi,\delta\psi)]dt$$

In this equation, by abuse of notation, $N(\psi, \delta\psi)$ is a linear combination of terms of the schematic form $\delta\psi\nabla^{-1}(\psi^2)$, $\delta\psi\nabla^{-1}(\psi\delta\psi)$ etc. Let's put $N(\psi, \delta\psi) = \delta\psi\nabla^{-1}(\psi^2)$, the other terms being treated along the same lines (but also requiring $\epsilon$ to be small enough). We note that

$$||P_k[\delta\psi\nabla^{-1}(\psi^2)]||_{L_t^B L_x^2} \leq CK2^{(1-\frac{1}{B})k}\tilde{c}_k,$$

where $B$ is an arbitrarily large number (the implied constants will depend on it). This follows from a simple frequency trichotomy and the bootstrap assumption. Now using Holder's inequality, we deduce

$$||\int_{T_1}^t P_k[N(\psi, \delta\psi)]dt||_{L_x^2} \leq C|t-T_1|^{1-\frac{1}{B}}2^{(1-\frac{1}{B})k}K\tilde{c}_k \leq C\frac{1}{C^{1-\frac{1}{B}}}K\tilde{c}_k$$

Clearly we also have

$$||\int_{T_1}^t \partial_i P_k\delta\psi_i dt||_{L_x^2} \leq C|t-T_1|2^k K\tilde{c}_k \leq C\frac{1}{C}K\tilde{c}_k.$$

Therefore, we infer that

$$||P_k\delta\psi_i||_{L_t^\infty L_x^2} \leq C(1+\frac{K}{C^{1-\frac{1}{B}}})\tilde{c}_k \leq C\frac{K}{100}\tilde{c}_k$$

provided $K$, $C$ are chosen[9] large enough (in relation to $B$). Arguing similarly, one deduces as well that

$$2^{\frac{k}{2}}||P_k\delta\psi_i||_{L_t^2L_x^2} + 2^{-\frac{k}{2}}||P_k\partial_t\delta\psi_i||_{L_t^2L_x^2} \leq C\frac{K}{C^{\frac{1}{2}}}\tilde{c}_k.$$

Using the fact that (see e. g. **[23]**)

$$||P_k\psi||_{S[k]([-T,T]\times\mathbf{R}^2)} \leq C||P_k\psi||_{L_t^\infty L_x^2([-T,T]\times\mathbf{R}^2)} + 2^{\frac{k}{2}}||P_k\psi||_{L_t^2L_x^2([-T,T]\times\mathbf{R}^2)}$$
$$+ 2^{-\frac{k}{2}}||P_k\partial_t\psi||_{L_t^2L_x^2([-T,T]\times\mathbf{R}^2)}$$

and choosing $K$, $C$ large enough, one deduces from this that

$$||P_k\delta\psi_i||_{S[k]([T_1,T]\times\mathbf{R}^2)} \leq C\frac{K}{2}\tilde{c}_k,$$

which is the desired conclusion for $P_k\delta\psi_i$. The argument for $\delta\psi_0$ is similar using (1.7). Thus we see that we may assume $|T-T_1| \geq \frac{2^{-k}}{C}$. Moreover, reiterating the preceding argument, and choosing $K$ large enough, we conclude that $||P_k\delta\psi_\nu||_{L_t^\infty L_x^2([T_1,T_1+\frac{2^{-k}}{C}]\times\mathbf{R}^2)} \leq \frac{K}{100C}\tilde{c}_k$. Now revert to the old notation

$$\Box\delta\psi_\nu = N(\delta\psi_\nu, \psi_\nu) = N_1(\delta\psi_\nu) + N_2(\delta\psi_\nu, \psi_\nu)$$

as in Proposition 2.2. Clearly, we have

$$||P_kN_1(\delta\psi_\nu, \psi_\nu)||_{N[k]([T_1,T]\times\mathbf{R}^2)} \leq ||P_k\chi(\frac{t}{T_0})N_1(\delta\psi_\nu, \psi_\nu)||_{N[k]([T_1,T]\times\mathbf{R}^2)}$$

$$||P_kN_2(\delta\psi_\nu, \psi_\nu)||_{N[k]([T_1,T]\times\mathbf{R}^2)} \leq C||P_kN_2(\delta\psi_\nu, \psi_\nu)||_{N[k]([-T,T]\times\mathbf{R}^2)}$$

Using 3.4(d) as well as time translation invariance, we can now infer that

$$||P_k\delta\psi_\nu||_{S[k]([T_1,T]\times\mathbf{R}^2)} \leq C\frac{K}{100}\tilde{c}_k + \zeta K\tilde{c}_k + \epsilon^2K^2\tilde{c}_k^2 \leq C\frac{K}{2}\tilde{c}_k$$

provided $\epsilon$, $\zeta$ are small enough. This yields the desired conclusion.

[9] Of course $C$ is chosen independently of $K$.

**2.0.7. The proof of theorem 1.11.** This is basically identical to the proof of theorem 1.10. Control over some subcritical norm $||u||_{L_t^\infty H^{1+\epsilon}}$ follows from standard Moser estimates instead of Corollary 1.1.

CHAPTER 3

# The Proof of Proposition 2.2

We have thus reduced the proof of theorem 1.10 to the verification of Proposition 2.2 in addition to the technical Moser type estimates allowing estimation of $||P_k\psi||_{S[k]}$. The proof of this Proposition is divided into the part dealing with expressions linear in $\delta\psi_\nu$, as well as those of higher degree of linearity. We commence by spelling out in detail the decomposition $N_\alpha(\delta\psi_\nu,\psi_\nu) = N_{1\alpha}(\delta\psi_\nu,\psi_\nu) + N_{2\alpha}(\delta\psi_\nu,\psi_\nu)$, where $\Box\delta\psi_\alpha = N_\alpha(\delta\psi_\nu,\psi_\nu)$. As in **[23]**, this decomposition requires extreme care in order to avoid too many time derivatives. Recalling (1.8), we define

$$N_1(\delta\psi_\nu,\psi_\nu) := \sum_{i=1}^{3} A_{i\alpha}(\delta\psi_\nu,\psi_\nu) + \sum_{i=1}^{5} B_{i\alpha}(\delta\psi_\nu,\psi_\nu) + \sum_{i=1}^{5} C_{i\alpha}$$

where

$$\begin{aligned} A_1(\delta\psi_\nu,\psi_\nu) =& i\partial^\beta[\delta\psi_\alpha\Delta^{-1}\sum_{j=1}^{2}\partial_j[R_\beta\psi^1R_j\psi^2 - R_\beta\psi^2R_j\psi^1]] \\ & - i\partial^\beta[\delta\psi_\beta\Delta^{-1}\sum_{j=1}^{2}\partial_j[R_\alpha\psi^1R_j\psi^2 - R_\alpha\psi^2R_j\psi^1]] \\ & + i\partial_\alpha[\delta\psi_\nu\Delta^{-1}\sum_{j=1}^{2}\partial_j[R^\nu\psi^1R_j\psi^2 - R^\nu\psi^2R_j\psi^1]] \end{aligned} \tag{3.1}$$

$$\begin{aligned} A_1(\delta\psi_\nu,\psi_\nu) =& + i\partial^\beta[\psi_\alpha\Delta^{-1}\sum_{j=1}^{2}\partial_j[R_\beta\delta\psi^1R_j\psi^2 - R_\beta\psi^2R_j\delta\psi^1]] \\ & - i\partial^\beta[\psi_\beta\Delta^{-1}\sum_{j=1}^{2}\partial_j[R_\alpha\delta\psi^1R_j\psi^2 - R_\alpha\psi^2R_j\delta\psi^1]] \\ & + i\partial_\alpha[\psi_\nu\Delta^{-1}\sum_{j=1}^{2}\partial_j[R^\nu\delta\psi^1R_j\psi^2 - R^\nu\psi^2R_j\delta\psi^1]] \end{aligned}$$

$$\begin{aligned} A_3(\delta\psi_\nu,\psi_\nu) =& + i\partial^\beta[\psi_\alpha\Delta^{-1}\sum_{j=1}^{2}\partial_j[R_\beta\psi^1R_j\delta\psi^2 - R_\beta\delta\psi^2R_j\psi^1]] \\ & - i\partial^\beta[\psi_\beta\Delta^{-1}\sum_{j=1}^{2}\partial_j[R_\alpha\psi^1R_j\delta\psi^2 - R_\alpha\delta\psi^2R_j\psi^1]] \\ & + i\partial_\alpha[\psi_\nu\Delta^{-1}\sum_{j=1}^{2}\partial_j[R^\nu\psi^1R_j\delta\psi^2 - R^\nu\delta\psi^2R_j\psi^1]]. \end{aligned}$$

$$B_1, C_1(\delta\psi_\nu, \psi_\nu) = \nabla_{x,t}[\delta\psi\nabla^{-1}[\psi\nabla^{-1}[\psi\nabla^{-1}(\psi^2)]]]$$
$$B_2, C_2(\delta\psi_\nu, \psi_\nu) = \nabla_{x,t}[\psi\nabla^{-1}[\delta\psi\nabla^{-1}[\psi\nabla^{-1}(\psi^2)]]]$$
$$B_3, C_3(\delta\psi_\nu, \psi_\nu) = \nabla_{x,t}[\psi\nabla^{-1}[\psi\nabla^{-1}[\delta\psi\nabla^{-1}(\psi^2)]]]$$
$$B_{4,5}, C_{4,5}(\delta\psi_\nu, \psi_\nu) = \nabla_{x,t}[\psi\nabla^{-1}[\psi\nabla^{-1}[\psi\nabla^{-1}(\delta\psi\psi)]]]$$

Of course we have used schematic notation for the $B$, $C$'s, as their fine structure won't matter. They are obtained by substituting one $\chi_\nu$ instead of the corresponding entry $\psi_\nu$ in the inner square bracket expressions on the right hand side of (1.8), where $\chi_\nu$ is the 'elliptic component' of the spherically symmetric $\psi_\nu$ in the decomposition $\psi_\nu = R_\nu\psi + \chi_\nu$. We recall $\psi_\nu = \psi_\nu^1 + i\psi_\nu^2$, $\delta\psi_\nu = \delta\psi_\nu^1 + i\delta\psi_\nu^2$, $\psi = -\sum_{j=1,2} R_j\psi_j$, $\delta\psi = -\sum_{j=1,2} R_j\delta\psi_j$. One can then define $N_{2\alpha}(\delta\psi_\nu, \psi_\nu) = N_\alpha(\delta\psi_\nu, \psi_\nu) - N_{1\alpha}(\delta\psi_\nu, \psi_\nu)$. The quintilinear terms above shall be relatively simple to estimate on account of the strong Strichartz type estimates satisfied by the $\psi_\nu$, see theorem 2.1 as well as the definition of $||.||_{S[k]}$. Unfortunately, the latter norm falls short of controlling $||.||_{L_t^2 L_x^\infty}$, which appears necessary in order to grant an elementary estimation of the trilinear terms $A_{i\alpha}$. We shall instead have to revert to the inherent null-structure in these terms as was done already in [**23**], in addition to the more complicated ingredients in $||.||_S$. The main new difficulty over the estimates in [**23**] has to do with the fact that we need to gain explicitly in time in these estimates. This would be relatively straightforward if we were working with Lebesgue type spaces; however, we shall work with null-frame spaces of type $L^2_{t_\omega} L^\infty_{x_\omega}$, which considerably complicates obtaining gains in time. The main novelty here(lemma 3.3) shall be a special type of decomposition of the spherical components $\psi_\nu$ into pieces which have well-defined physical as well as frequency localization properties. More precisely, we shall be able to physically localize $\psi_\nu$ closely to the light cone. This part will then be written as a sum of two components, the first of which can be written as a sum of pieces which propagate in a direction essentially opposite to their physical support. Thus the first component is obtained by first localizing $\psi_\nu$ to an angular sector in Fourier space, then multiplying with a physical cutoff localizing to an *opposite or identical*[1] angular sector, and finally summing over all sectors. The size of the angular sectors shall essentially be dictated by the $\zeta$ in the statement of Proposition 2.2. While the first component is exactly the part which fails to decay in $L^2_{t_\omega} L^\infty_{x_\omega}$ as $t \to \infty$, it does lead to improved trilinear null-form estimates due to the dual localization properties. The 2nd component in turn will decay like a standard Lebesgue norm as $t \to \infty$. The next subsection contains the core estimates. As the estimates are rather technical, we briefly explain the strategy of the proof, which is conceptually simple:

**(1)**: First, upon localizing the nonlinearity to a time interval $t \sim 2^i$, one tries to reduce the frequencies of all functions occuring inside the nonlinearity to absolute size $<< 2^{\delta i}$, for some small $\delta > 0$. The idea here is that far apart frequencies should interact little. But this in addition to the refined control over the frequency modes of the spherically symmetric components should suffice to get control over the cases when extremely small or large frequencies are present. The tool to achieve this are the refined trilinear estimates in 3.4(c). Unfortunately, these estimates aren't quite good enough to control certain high-high interactions, which accounts for a number

[1]This depends on whether the space-time Fourier support is contained in the upper half-space $\tau > 0$ or lower half-space $\tau < 0$.

of extra cases that need to be considered.

**(2)** Having controlled the cases when the frequencies are very small or large in relation to the time interval one works on, one now tries to exploit the pointwise estimates provided by Christodoulou-Tahvildar-Zadeh, since one has gained some room to lose in the frequencies. The device here is the decomposition of the spherically symmetric components referred to in the preceding paragraph, which is a direct consequence of the pointwise decay estimates. This allows one to decompose these components into pieces that disperse quickly enough, as well as other pieces that interact very weakly. Of course one exploits the trilinear structure of the nonlinearity to make this work.

**3.0.8. Estimating the trilinear null-forms.** We use the operator $I=\sum_{k\in\mathbf{Z}}P_kQ_{<k+100}$ as before and employ the schematic decomposition

$$\nabla_{x,t}[\psi_1\nabla^{-1}[\psi_2\psi_3]]=\nabla_{x,t}[\psi_1 I\nabla^{-1}[\psi_2\psi_3]]+\nabla_{x,t}[\psi_1(1-I)\nabla^{-1}[\psi_2\psi_3]]$$

for each of the $A_{i\alpha}$'s. In order to make sense of this, one needs to substitute Schwartz extensions for the inputs $\delta\psi^{1,2}|_{[T_1,T]}$, $\psi^{1,2}|_{[T_1,T]}$ of the inner square brackets, in accordance with the bootstrap assumption in Proposition 2.2. In the following we shall localize the frequency localized nonlinearities $P_kN(...)$ to a dyadic time interval $t\sim 2^i$ and strive for an estimate of the form

$$||\chi_i(t)P_kN(...)||_{N[k]}\leq C2^{-\mu i}\tilde{c}_k$$

One can then sum over $i$ large enough to obtain the estimate in Proposition 2.1.

**(A): The large modulation case.** Estimating the terms

$$(I):\ \nabla_{x,t}[\delta\psi\triangle^{-1}\sum_{j=1,2}\partial_j(1-I)[R_\beta\psi_2R_j\psi_3-R_j\psi_2R_\beta\psi_3]]$$

$$(II):\ \nabla_{x,t}[\psi\triangle^{-1}\sum_{j=1,2}\partial_j(1-I)[R_\beta\delta\psi_2R_j\psi_3-R_j\delta\psi_2R_\beta\psi_3]]$$

(I): *The first term.* We use the decomposition

$$\begin{aligned}&\chi(\frac{t}{\tilde{T}_0})\nabla_{x,t}[\delta\psi\triangle^{-1}\sum_{j=1,2}\partial_j(1-I)[R_\beta\psi_2R_j\psi_3-R_j\psi_2R_\beta\psi_3]]\\&=\sum_{i\geq\log_2\tilde{T}_0}\chi_i(t)\nabla_{x,t}[\delta\psi\triangle^{-1}\sum_{j=1,2}\partial_j(1-I)[R_\beta\psi_2R_j\psi_3-R_j\psi_2R_\beta\psi_3]],\end{aligned}$$

where $\chi_i(t)$ smoothly localizes to the interval $t\sim 2^i$. Then we localize the frequencies and freeze $i\in\mathbf{Z}$, arriving at an expression

$$\chi_i(t)\nabla_{x,t}P_{k_0}[P_{k_1}\delta\psi\triangle^{-1}\sum_{j=1,2}\partial_j(1-I)P_k[R_\beta P_{k_2}\psi_2R_jP_{k_3}\psi_3-R_jP_{k_2}\psi_2R_\beta P_{k_3}\psi_3]]$$

We distinguish between the following cases:
(I.a): *One of the following options hold:* $i\leq C|k_2|$, $i\leq C|k_3|$, $i\leq C|k_0-k_1|$, $i\leq C\min\{|k-k_1|,|k-k_2|\}$. This case is handled by means of lemma 2.6 as well as the following lemma, provided $P_{k_{2,3}}\psi_{2,3}$ are of the first type:

LEMMA 3.1. [23] *Let* $\psi_{1,2,3} \in \mathbf{R}^{2+1}$. *Then, for integers* $k_{1,2,3}$ *and suitable* $\delta_{1,2} > 0$, *the following inequality holds:*

$$||\nabla_{x,t}P_0[P_{k_1}\psi_1\nabla^{-1}(1-I)[R_\nu P_{k_2}\psi_2 R_j P_{k_3}\psi_3 - R_j P_{k_2}\psi_2 R_\nu P_{k_3}\psi_3]]||_{N[0]}$$
$$\leq C 2^{\delta_1 \min\{-\min\{k_1,k_2,k_3\},0\}} \prod_i 2^{\delta_2 \min\{\max_{j\neq i}\{k_i, k_i - k_j\},0\}} \prod_l ||P_{k_l}\psi_l||_{S[k_l]}$$

Indeed, observe that if $i \leq C\max\{|k_2|,|k_3|\}$, we obtain from lemma 2.6 that $\min\{||P_{k_2}\psi_2||_{S[k_2]}, ||P_{k_3}\psi_3||_{S[k_3]}\} \leq C2^{-\mu i}$. Carrying out the summations over $k$, $k_i$ satisfying these assumptions, we arrive at the upper bound $\leq C2^{-\mu i}K\tilde{c}_{k_0}$. Summing over $i \geq \log_2 \tilde{T}_0$ results in a small exponential gain in $T_0 \leq \tilde{T}_0$. If one of the other cases occurs, one gets an exponential gain $2^{-\min\{\delta_1,\delta_2\}i}$ from the above lemma. We are fudging a bit since we have thrown the localizer $\chi_i(t)$ in front, and this may affect the space-time Fourier support of the expression, hence its norm $||.||_{N[k_0]}$. However, this is detrimental only if the modulation (i. e. distance of the space-time Fourier support to the light cone) is $\leq C2^{-i}$, and only affects those parts estimated with respect to $||.||_{\dot{X}_{k_0}^{-1,-\frac{1}{2},1}}$, as null-frame spaces aren't needed yet, see the proof in [23]. Assuming $Q_{<-i+O(1)}(\text{Output})$ to be a $\dot{X}_{k_0}^{-1,-\frac{1}{2},1}$-atom, we estimate

$$||\chi_i(t)P_{k_0}Q_{<-i+O(1)}(\text{Output})||_{L_t^1\dot{H}^{-1}} \leq C||\chi_i(t)||_{L_t^2L_x^\infty}||P_{k_0}Q_{<-i+O(1)}(\text{Output})||_{L_t^2\dot{H}^{-1}}$$
$$\leq C\sum_{a<0} 2^{\frac{i}{2}}2^{\frac{-i+a}{2}} \leq C1$$

Thus the cutoff is irrelevant.
Now assume at least one of $P_{k_{2,3}}$ is of the 2nd type. We need the following lemma

LEMMA 3.2. *Let* $\psi_{2,3} \in \mathcal{S}(\mathbf{R}^{2+1})$. *Assume also that* $||P_{k_{2,3}}\psi_{2,3}||_{S[k_{2,3}]} \leq C\frac{\tilde{c}_{k_{2,3}}}{\epsilon}$ *with a frequency envelope* $\tilde{c}_k$ *as in the preceding. Assume that* $P_{k_2}\psi_2$ *is of the 2nd type, and* $P_{k_3}\psi_3$ *admits a decomposition into functions of first and 2nd type as enunciated in theorem 2.3. Then we have for suitable* $\delta_{1,2} > 0$

$$||P_k[R_\beta P_{k_2}\psi_2 R_j P_{k_3}\psi_3 - R_j P_{k_2}\psi_2 R_\beta P_{k_3}\psi_3]||_{L_t^2L_x^2}$$
$$\leq C2^{\frac{\min\{k_3,k\}}{2}} 2^{\delta_1 \min\{k_2-k_3,0\}} 2^{\delta_2 \min\{k-k_2\}} [\frac{\tilde{c}_{k_2}}{\epsilon} + \frac{\tilde{c}_{k_3}}{\epsilon}]$$

PROOF. First assume that $P_{k_3}\psi_3$ is of the first type. Using the definition of $||.||_{S[k]}$, we infer the desired estimate for the contributions of

$$P_k[R_\beta P_{k_2}(1-I)\psi_2 R_j P_{k_3}\psi_3 - R_j P_{k_2}(1-I)\psi_2 R_\beta P_{k_3}\psi_3]$$

and similarly for

$$P_k[R_\beta P_{k_2}\psi_2 R_j P_{k_3}(1-I)\psi_3 - R_j P_{k_2}\psi_2 R_\beta(1-I) P_{k_3}\psi_3]$$

Take the first expression: first consider the case $k_2 = k_3 + O(1)$. We estimate, using theorem 2.3

$$||P_k[R_\beta P_{k_2}(1-I)\psi_2 R_j P_{k_3}\psi_3 - R_j P_{k_2}(1-I)\psi_2 R_\beta P_{k_3}\psi_3]||_{L_t^2L_x^2}$$
$$\leq C2^{2k(\frac{1}{1+\mu}-\frac{1}{2})}||P_k[R_\beta P_{k_2}(1-I)\psi_2 R_j P_{k_3}\psi_3 - R_j P_{k_2}(1-I)\psi_2 R_\beta P_{k_3}\psi_3]||_{L_t^2L_x^{1+\mu}}$$
$$\leq C2^{\frac{k}{2}}2^{\frac{(k-k_2)(1-3\mu)}{2+2\mu}}\frac{\tilde{c}_{k_3}}{\epsilon}$$

Next, in case $k_2 << k_3$, we estimate

$$||P_k[R_\beta P_{k_2}(1-I)\psi_2 R_j P_{k_3}\psi_3]||_{L_t^2 L_x^2}$$
$$\leq C(\sum_{c\in C_{k_3,k_2-k_3}} ||P_k[R_\beta P_{k_2}(1-I)\psi_2 R_j P_c\psi_3]||^2_{L_t^2 L_x^2})^{\frac{1}{2}} \leq C2^{\frac{k_3}{2}} 2^{\delta_1(k_2-k_3)}\frac{\tilde{c}_{k_3}}{\epsilon}$$

The remaining term is estimated similarly, as is the case when $k_2 >> k_3$. Further, if for example $k_2 = k_3 + O(1)$, we can estimate

$$||P_k[R_\beta P_{k_2} I\psi_2 R_j P_{k_3}\psi_3 - R_j P_{k_2} I\psi_2 R_\beta P_{k_3}\psi_3]||_{L_t^2 L_x^2}$$
$$\leq C2^{(1-\epsilon)k}||P_{k_2}\psi_2||_{L_t^2 L_x^{2+}}||P_{k_3}\psi_3||_{L_t^\infty L_x^2} \leq C2^{(1-\epsilon)k-(\frac{1}{2}-\epsilon)k_2}\frac{\tilde{c}_{k_2}}{\epsilon}$$

The remaining cases $k = k_2 + O(1)$, $k = k_3 + O(1)$ are handled similarly. Now assume that both $P_{k_2}\psi_2$ and $P_{k_3}\psi_3$ are of 2nd type. In that case, if $k_1 = k_2 + O(1)$, estimate

$$||P_k[R_\beta P_{k_1}\psi_1 P_{k_2} R_j\psi_2]||_{L_t^2 L_x^2} \leq C2^{(1-\epsilon)k}||R_\beta P_{k_1}\psi_1||_{L_t^\infty L_x^2}||P_{k_2}\psi_2||_{L_t^2 L_x^{2+}}$$
$$\leq C2^{(1-\epsilon)(k-k_2)}2^{\frac{k}{2}}\frac{\tilde{c}_{k_2}}{\epsilon}$$

The remaining frequency interactions are treated similarly. □

Returning to case (I.a) when at least one of $P_{k_{2,3}}\psi_{2,3}$ is of 2nd type, we claim that we have the estimate

$$||\nabla_{x,t}P_{k_0}[P_{k_1}\delta\psi_1\nabla^{-1}(1-I)P_k[R_\nu P_{k_2}\psi_2 R_j P_{k_3}\psi_3 - R_j P_{k_2}\psi_2 R_\nu P_{k_3}\psi_3]]||_{N[k_0]}$$
$$\leq C2^{\delta_1[\min_{i=2,3}\{k,k_i\}-\max_{i=2,3}\{k,k_i\}]}2^{-\delta_2(|k-k_1|)}[\frac{\tilde{c}_{k_2}}{\epsilon}+\frac{\tilde{c}_{k_3}}{\epsilon}]\tilde{c}_{k_0}$$

One could then sum over all frequency parameters (except $k_0$) and obtain the required exponential gain in $i$ under the hypotheses of case (I.a)[2]. To verify this estimate, we may assume $k_0 = 0$. One needs to distinguish between $k_1 \in [-10, 10]$, $k_1 > 10$, $k_1 < -10$. These are similar, so we treat the first case: we have

$$||\nabla_{x,t}P_0 Q_{>10}[P_{k_1}\delta\psi_1\nabla^{-1}(1-I)P_k[R_\nu P_{k_2}\psi_2 R_j P_{k_3}\psi_3 - R_j P_{k_2}\psi_2 R_\nu P_{k_3}\psi_3]]||_{N[0]}$$
$$\leq C||P_0 Q_{>10}[P_{k_1}\delta\psi_1\nabla^{-1}(1-I)P_k[R_\nu P_{k_2}\psi_2 R_j P_{k_3}\psi_3 - R_j P_{k_2}\psi_2 R_\nu P_{k_3}\psi_3]]||_{L_t^2 L_x^2}$$
$$\leq C||P_{k_1}\delta\psi_1||_{L_t^\infty L_x^2}2^{\frac{\min\{k,k_3\}}{2}}2^{\delta_1\min\{k_2-k_3,0\}}2^{\delta_2\min\{k-k_2,0\}}[\frac{\tilde{c}_{k_2}}{\epsilon}+\frac{\tilde{c}_{k_3}}{\epsilon}]\tilde{c}_0$$

One checks that this verifies the claim, with a lot to spare. Next, we can estimate

$$||\nabla_{x,t}P_0 Q_{<10}[P_{k_1}Q_{<k-100}\delta\psi_1$$
$$\nabla^{-1}(1-I)P_k[R_\nu P_{k_2}\psi_2 R_j P_{k_3}\psi_3 - R_j P_{k_2}\psi_2 R_\nu P_{k_3}\psi_3]]||_{N[0]}$$
$$\leq C||\nabla_{x,t}P_0 Q_{<10}[P_{k_1}Q_{<k-100}\delta\psi_1$$
$$\nabla^{-1}(1-I)P_k[R_\nu P_{k_2}\psi_2 R_j P_{k_3}\psi_3 - R_j P_{k_2}\psi_2 R_\nu P_{k_3}\psi_3]]||_{\dot{X}_0^{-1,-\frac{1}{2},1}}$$
$$\leq C2^{-\frac{k}{2}}2^{\frac{\min\{k,k_3\}}{2}}2^{\delta_1\min\{k_2-k_3,0\}}2^{\delta_2\min\{k-k_2,0\}}\tilde{c}_0[\frac{\tilde{c}_{k_2}}{\epsilon}+\frac{\tilde{c}_{k_3}}{\epsilon}]$$

[2]The cutoff $\chi_i(t)$ in front is handled as before.

Again this verifies the claim. Finally, we have the estimate

$$\begin{aligned}
&||\nabla_{x,t}P_0 Q_{<10}[P_{k_1}Q_{\geq k-100}\delta\psi_1\\
&\qquad\qquad \nabla^{-1}(1-I)P_k[R_\nu P_{k_2}\psi_2 R_j P_{k_3}\psi_3 - R_j P_{k_2}\psi_2 R_\nu P_{k_3}\psi_3]]||_{N[0]}\\
&\leq C||\nabla_{x,t}P_0 Q_{<10}[P_{k_1}Q_{\geq k-100}\delta\psi_1\\
&\qquad\qquad \nabla^{-1}(1-I)P_k[R_\nu P_{k_2}\psi_2 R_j P_{k_3}\psi_3 - R_j P_{k_2}\psi_2 R_\nu P_{k_3}\psi_3]]||_{L_t^1\dot{H}^{-1}}\\
&\leq C||P_{k_1}Q_{\geq k-100}\delta\psi_1||_{L_t^2L_x^2}\\
&\qquad\qquad ||\nabla^{-1}(1-I)P_k[R_\nu P_{k_2}\psi_2 R_j P_{k_3}\psi_3 - R_j P_{k_2}\psi_2 R_\nu P_{k_3}\psi_3]||_{L_t^2L_x^\infty}\\
&\leq C2^{-\frac{k}{2}}2^{\frac{\min\{k,k_3\}}{2}}2^{\delta_1\min\{k_2-k_3,0\}}2^{\delta_2\min\{k-k_2,0\}}\tilde{c}_0[\frac{\tilde{c}_{k_2}}{\epsilon}+\frac{\tilde{c}_{k_3}}{\epsilon}]
\end{aligned}$$

as in the preceding estimate. This concludes case (I.a).
(I.b): *$i \leq C|k|$, $i \leq C|k_1|$, and none of the properties in (I.a) hold.* Thus in this case, we have $k \leq C-i$, $k_1 \leq C-i$, $|k-k_1| << i$; we may treat the last difference as $O(1)$[3]. In this case we have to work harder to obtain the exponential gain in $i$, since the previous trilinear estimates won't suffice. Observe that we only need to worry about the case $\nu = 0$, though, since otherwise one can pull a derivative out of the inner square bracket expression. Also, we may easily reduce the Fourier support of $P_{k_{2,3}}\psi_{2,3}$ to the hyperbolic regime[4] (distance to light cone at most comparable to frequency). First, consider the case $|k| > (1+\mu)i$, for some small $\mu > 0$. In that case, we have

$$\begin{aligned}
&||\nabla_{x,t}P_{k_0}\chi_i(t)[P_{k_1}\delta\psi_1\nabla^{-1}(1-I)P_k[R_\nu P_{k_2}\psi_2 R_j P_{k_3}\psi_3 - R_j P_{k_2}\psi_2 R_\nu P_{k_3}\psi_3]]||_{N[k_0]}\\
&\leq C||\nabla_{x,t}P_{k_0}\chi_i(t)[P_{k_1}Q_{\geq k-100}\delta\psi_1\\
&\qquad\qquad \nabla^{-1}(1-I)P_k[R_\nu P_{k_2}\psi_2 R_j P_{k_3}\psi_3 - R_j P_{k_2}\psi_2 R_\nu P_{k_3}\psi_3]]||_{N[k_0]}\\
&+||\nabla_{x,t}P_{k_0}\chi_i(t)[P_{k_1}Q_{<k-100}\delta\psi_1\\
&\qquad\qquad \nabla^{-1}(1-I)P_k[R_\nu P_{k_2}\psi_2 R_j P_{k_3}\psi_3 - R_j P_{k_2}\psi_2 R_\nu P_{k_3}\psi_3]]||_{N[k_0]}
\end{aligned}$$

The first summand is further decomposed as follows:

$$\begin{aligned}
&||\nabla_{x,t}P_{k_0}\chi_i(t)[P_{k_1}Q_{\geq k-100}\delta\psi_1\\
&\qquad\qquad \nabla^{-1}(1-I)P_k[R_\nu P_{k_2}\psi_2 R_j P_{k_3}\psi_3 - R_j P_{k_2}\psi_2 R_\nu P_{k_3}\psi_3]]||_{N[k_0]}\\
&\leq C||\nabla_{x,t}P_{k_0}Q_{>k_0}\chi_i(t)[P_{k_1}Q_{\geq k-100}\delta\psi_1\\
&\qquad\qquad \nabla^{-1}(1-I)P_k[R_\nu P_{k_2}\psi_2 R_j P_{k_3}\psi_3 - R_j P_{k_2}\psi_2 R_\nu P_{k_3}\psi_3]]||_{\dot{X}_{k_0}^{-\frac{1}{2},-1,2}}\\
&+||\nabla_{x,t}P_{k_0}Q_{<k_0}\chi_i(t)[P_{k_1}Q_{\geq k-100}\delta\psi_1\\
&\qquad\qquad \nabla^{-1}(1-I)P_k[R_\nu P_{k_2}\psi_2 R_j P_{k_3}\psi_3 - R_j P_{k_2}\psi_2 R_\nu P_{k_3}\psi_3]]||_{L_t^1\dot{H}^{-1}}
\end{aligned}$$

---

[3]We do this in order to avoid carrying too many small constants around; this is legitimate since the exponential gains obtained later are independent.

[4]We shall not include the localizers $Q_{<k_{2,3}}$ everywhere in order to streamline notation.

We then estimate

$$\begin{aligned}&||\nabla_{x,t}P_{k_0}Q_{>k_0}\chi_i(t)[P_{k_1}Q_{\geq k-100}\delta\psi_1\\&\qquad\nabla^{-1}(1-I)P_k[R_\nu P_{k_2}\psi_2R_jP_{k_3}\psi_3-R_jP_{k_2}\psi_2R_\nu P_{k_3}\psi_3]]||_{\dot{X}_{k_0}^{-\frac{1}{2},-1,2}}\\&\leq C2^{\frac{k_0}{2}}||\chi_i(t)||_{L_t^2}||P_{k_1}Q_{\geq k-100}\delta\psi_1||_{L_t^\infty L_x^2}\\&\qquad||\nabla^{-1}(1-I)P_k[R_\nu P_{k_2}\psi_2R_jP_{k_3}\psi_3-R_jP_{k_2}\psi_2R_\nu P_{k_3}\psi_3]||_{L_t^\infty L_x^2}\\&\leq C2^{\frac{k_0}{2}+\frac{i}{2}}\tilde{c}_{k_1}\leq C2^{-\frac{\mu}{2}i}\tilde{c}_{k_1}\end{aligned}$$

Similarly, we have

$$\begin{aligned}&||\nabla_{x,t}P_{k_0}Q_{<k_0}\chi_i(t)[P_{k_1}Q_{\geq k-100}\delta\psi_1\\&\qquad\nabla^{-1}(1-I)P_k[R_\nu P_{k_2}\psi_2R_jP_{k_3}\psi_3-R_jP_{k_2}\psi_2R_\nu P_{k_3}\psi_3]]||_{L_t^1\dot{H}^{-1}}\\&\leq C2^{k_0}||\chi_i(t)||_{L_t^2}||P_{k_1}Q_{\geq k-100}\delta\psi_1||_{L_t^2L_x^2}\\&\qquad||\nabla^{-1}(1-I)P_k[R_\nu P_{k_2}\psi_2R_jP_{k_3}\psi_3-R_jP_{k_2}\psi_2R_\nu P_{k_3}\psi_3]||_{L_t^\infty L_x^2},\end{aligned}$$

and this is controlled by $2^{k_0-\frac{k}{2}+\frac{i}{2}}\tilde{c}_{k_1}\leq C2^{-\frac{\mu}{2}i}\tilde{c}_{k_1}$ as desired. The remaining terms are handled similarly. Thus we now assume that $i\leq C|k|\leq(1+\mu)|i|$. We then claim that we may replace the operator $(1-I)$ by $Q_{>\frac{k}{2}}$. Indeed, we have

$$\begin{aligned}&||\nabla_{x,t}P_{k_0}\chi_i(t)[P_{k_1}\delta\psi_1\\&\qquad\nabla^{-1}(1-I)P_kQ_{<\frac{k}{2}}[R_\nu P_{k_2}\psi_2R_jP_{k_3}\psi_3-R_jP_{k_2}\psi_2R_\nu P_{k_3}\psi_3]]||_{N[k_0]}\\&\leq C\sum_{k+100<a<\frac{k}{2}}||\nabla_{x,t}P_{k_0}\chi_i(t)[P_{k_1}Q_{<a-100}\delta\psi_1\\&\qquad\nabla^{-1}(1-I)P_kQ_a[R_\nu P_{k_2}\psi_2R_jP_{k_3}\psi_3-R_jP_{k_2}\psi_2R_\nu P_{k_3}\psi_3]]||_{N[k_0]}\\&\quad+\sum_{k+100<a<\frac{k}{2}}||\nabla_{x,t}P_{k_0}\chi_i(t)[P_{k_1}Q_{\geq a-100}\delta\psi_1\\&\qquad\nabla^{-1}(1-I)P_kQ_a[R_\nu P_{k_2}\psi_2R_jP_{k_3}\psi_3-R_jP_{k_2}\psi_2R_\nu P_{k_3}\psi_3]]||_{N[k_0]}\end{aligned}$$

We treat the first summand, the 2nd being similar. We have

$$\begin{aligned}&||\nabla_{x,t}P_{k_0}[\chi_i(t)[P_{k_1}Q_{<a-100}\delta\psi_1\\&\qquad\nabla^{-1}(1-I)P_kQ_a[R_\nu P_{k_2}\psi_2R_jP_{k_3}\psi_3-R_jP_{k_2}\psi_2R_\nu P_{k_3}\psi_3]]]||_{N[k_0]}\\&\leq C2^{-\frac{k_0}{2}}||P_{k_0}Q_{a+O(1)}[\chi_i(t)[P_{k_1}Q_{<a-100}\delta\psi_1\\&\qquad\nabla^{-1}(1-I)P_kQ_a[R_\nu P_{k_2}\psi_2R_jP_{k_3}\psi_3-R_jP_{k_2}\psi_2R_\nu P_{k_3}\psi_3]]]||_{L_t^2L_x^2}\\&\leq C2^{\frac{a+k}{2}}\tilde{c}_{k_1}\end{aligned}$$

This is clearly acceptable. We now notice the identity

$$-2(\partial_t\psi_2\partial_r\psi_3-\partial_r\psi_2\partial_t\psi_3)=(\partial_t+\partial_r)\psi_2(\partial_t-\partial_r)\psi_3-(\partial_t-\partial_r)\psi_2(\partial_t+\partial_r)\psi_3$$

Applying this to our frequency localized situation, we have the identity[5](recall that $\psi_{1,2}$ are radial)

$$\begin{aligned}&R_0P_{k_2}Q_{<k_2}\psi_2R_iP_{k_3}Q_{<k_3}\psi_3 - R_iP_{k_2}Q_{<k_2}\psi_2R_0P_{k_3}Q_{<k_3}\psi_3\\&=\frac{x_i}{r}[(\partial_t+\partial_r)\nabla^{-1}P_{k_2}Q_{<k_2}\psi_2(\partial_t-\partial_r)\nabla^{-1}P_{k_3}Q_{<k_3}\psi_3\\&-(\partial_t-\partial_r)\nabla^{-1}P_{k_2}Q_{<k_2}\psi_2(\partial_t+\partial_r)\nabla^{-1}P_{k_3}Q_{<k_3}\psi_3]\end{aligned}$$

Now let $\phi\in C_0^\infty(\mathbf{R})$ be a smooth cutoff and use the decomposition $\psi_{2,3}=\phi_{2^{-\frac{i}{2+}}}(u)\psi_{2,3}+(1-\phi_{2^{-\frac{i}{2+}}}(u))\psi_{2,3}$, where $u=t-r$ and $\phi_\lambda(u)=\phi(\frac{u}{\lambda})$. Now Proposition 1.1 implies that

$$||\chi_i(t)\phi_{2^{-\frac{i}{2+}}}(u)\psi_{2,3}||_{L_x^2}\leq C\sqrt{t^{-1}\times t\times 2^{-\frac{i}{2+}}}\leq C2^{-\frac{i}{4+}}$$

Now let $\frac{2}{4+}+\frac{1}{M}=\frac{1}{2}$ and return to the full expression. We first estimate the large-modulation contribution:

$$\begin{aligned}&||P_{k_0}Q_{>k_0}\nabla_{x,t}\chi_i(t)[P_{k_1}\delta\psi_1\\&\nabla^{-1}P_kQ_{>\frac{k}{2}}[R_0P_{k_2}Q_{<k_2}[\phi_{2^{-\frac{i}{2+}}}(u)\psi_2]R_iP_{k_3}Q_{<k_3}\psi_3\\&\qquad-R_iP_{k_2}Q_{<k_2}[\phi_{2^{-\frac{i}{2+}}}(u)\psi_2]R_0P_{k_3}Q_{<k_3}\psi_3||_{N[k_0]}\\&\leq C||P_{k_0}Q_{>k_0}\nabla_{x,t}\chi_i(t)[P_{k_1}\delta\psi_1\\&\qquad\nabla^{-1}P_kQ_{\geq\frac{k}{2}}[R_0P_{k_2}Q_{<k_2}[\phi_{2^{-\frac{i}{2+}}}(u)\psi_2]R_iP_{k_3}Q_{<k_3}\psi_3\\&\qquad-R_iP_{k_2}Q_{<k_2}[\phi_{2^{-\frac{i}{2+}}}(u)\psi_2]R_0P_{k_3}Q_{<k_3}\psi_3||_{\dot{X}_{k_0}^{-\frac{1}{2},-1,2}}\\&\leq C2^{-\frac{k_0}{2}}2^{(1-\epsilon)k_0-k_1}||\chi_i(t)||_{L_t^{4+}}||P_{k_1}\delta\psi_1||_{L_t^{4+}L_x^\infty}\\&\qquad||R_0P_{k_2}Q_{<k_2}[\phi_{2^{-\frac{i}{2+}}}(u)\psi_2]||_{L_t^\infty L_x^2}||P_{k_3}\psi_3||_{L_t^ML_x^{2+}}\\&\leq C2^{(\frac{1}{2}-\epsilon)k_0-\frac{k_1}{4+}}\tilde{c}_{k_1}\frac{\tilde{c}_{k_3}}{\epsilon}\end{aligned}$$

Keeping our assumptions on the frequencies in mind, this is more than what we need. Next, restricting the expression to modulation $\leq 2^{k_0}$, we have

$$\begin{aligned}&||P_{k_0}Q_{\leq k_0}\nabla_{x,t}\chi_i(t)[P_{k_1}\delta\psi_1\\&\nabla^{-1}P_kQ_{\geq\frac{k}{2}}[R_0P_{k_2}Q_{<k_2}[\phi_{2^{-\frac{i}{2+}}}(u)\psi_2]R_iP_{k_3}Q_{<k_3}\psi_3\\&\qquad-R_iP_{k_2}Q_{<k_2}[\phi_{2^{-\frac{i}{2+}}}(u)\psi_2]R_0P_{k_3}Q_{<k_3}\psi_3||_{N[k_0]}\\&\leq C\sum_{k_2+O(1)>a\geq\frac{k}{2}}||P_{k_0}Q_{\leq k_0}\nabla_{x,t}\chi_i(t)[P_{k_1}Q_{a+O(1)}\delta\psi_1\\&\nabla^{-1}P_kQ_a[R_0P_{k_2}Q_{<k_2}[\phi_{2^{-\frac{i}{2+}}}(u)\psi_2]R_iP_{k_3}Q_{<k_3}\psi_3\\&\qquad-R_iP_{k_2}Q_{<k_2}[\phi_{2^{-\frac{i}{2+}}}(u)\psi_2]R_0P_{k_3}Q_{<k_3}\psi_3||_{L_t^1\dot{H}^{-1}}\\&\leq C\sum_{k_2+O(1)>a>\frac{k}{2}}2^{a-k}2^{\frac{k}{2+}}||\chi_i(t)||_{L_t^{4+}}||P_{k_1}Q_{a+O(1)}\delta\psi_1||_{L_t^2L_x^\infty}\\&\qquad||\phi_{2^{-\frac{i}{2+}}}(u)\psi_2||_{L_t^\infty L_x^2}||P_{k_3}Q_{<k_3}\psi_3||_{L_t^{4-}L_x^{4+}}\end{aligned}$$

[5]Recall the suppressed localizations, see previous footnote.

We can bound this by

$$2^{a-k}2^{\frac{i}{4+}}2^{\frac{k}{2+}}2^{k_1}2^{-\frac{i}{4+}}2^{-\frac{a}{2}}\tilde{c}_{k_1}$$

Our assumptions ensure that we may sum over $\frac{k}{2} < a < k_2 + O(1)$, resulting in an exponential gain in $i$. An identical argument may be used when $P_{k_3}Q_{<k_3}\psi_3$ is replaced by $P_{k_3}Q_{<k_3}[\phi_{2^{-\frac{i}{2+}}}(u)\psi_3]$, so we may replace both $P_{k_{2,3}}Q_{<k_{2,3}}\psi_{2,3}$ by $P_{k_{2,3}}Q_{<k_{2,3}}[(1-\phi_{2^{-\frac{i}{2+}}}(u))\psi_3]$. In that case we utilize the null-form identity recorded earlier: use

$$\begin{aligned}
&R_0P_{k_2}Q_{<k_2}[(1-\phi_{2^{-\frac{i}{2+}}})\psi_2]R_iP_{k_3}Q_{<k_3}[(1-\phi_{2^{-\frac{i}{2+}}})\psi_3]\\
&\qquad - R_iP_{k_2}Q_{<k_2}[(1-\phi_{2^{-\frac{i}{2+}}})\psi_2]R_0P_{k_3}Q_{<k_3}[(1-\phi_{2^{-\frac{i}{2+}}})\psi_3]\\
&= \frac{x_i}{r}[(\partial_t+\partial_r)\nabla^{-1}P_{k_2}Q_{<k_2}[(1-\phi_{2^{-\frac{i}{2+}}})\psi_2](\partial_t-\partial_r)\nabla^{-1}P_{k_3}Q_{<k_3}[(1-\phi_{2^{-\frac{i}{2+}}})\psi_3]\\
&- \frac{x_i}{r}[(\partial_t-\partial_r)\nabla^{-1}P_{k_2}Q_{<k_2}[(1-\phi_{2^{-\frac{i}{2+}}})\psi_2](\partial_t+\partial_r)\nabla^{-1}P_{k_3}Q_{<k_3}[(1-\phi_{2^{-\frac{i}{2+}}})\psi_3]
\end{aligned}$$

We can now exploit the fact that $|\frac{k}{2}| < \frac{(1+\mu)i}{2} < i(1-\delta)$ for $\mu$ small enough, as well as $T_0$ large enough. Thus we may move the multiplier $\chi_i(t)$ past the Fourier multiplier $\nabla^{-1}P_kQ_{>\frac{k}{2}}$ while trading in errors exponentially decreasing [6]outside of $t \sim 2^i$. In other words, under the present assumptions on the frequencies, we may write schematically

$$\begin{aligned}
&\chi_i(t)\nabla_{x,t}P_{k_0}[P_{k_1}\delta\psi_1\nabla^{-1}P_kQ_{>\frac{k}{2}}[,]]\\
&\qquad = \chi_i(t)\nabla_{x,t}P_{k_0}[P_{k_1}\delta\psi_1\nabla^{-1}P_kQ_{>\frac{k}{2}}(\chi_{i1}(t)+\chi_{i2}(t))[,]],
\end{aligned}$$

where $\chi_{i1}(t)$ is supported on $t \sim 2^i$ while $|\chi_{i2}(t)| \leq C2^{-Ni}$ for $t < 2^{i+O(1)}$, as well as $|\chi_{i2}(t)| < t^{-N}$ for $t >> 2^i$. It is then easy to verify that this leads to acceptable terms, so we may focus on the contribution of $\chi_{1i}(t)$. We shall want to move the operator $\partial_t + \partial_r$ past the Fourier localizer $P_{k_2}Q_{<k_2}$. We write

$$\begin{aligned}
&P_{k_2}Q_{<k_2}[(1-\phi_{2^{-\frac{i}{2+}}}(u))\psi_2]\\
&= \rho_i(r)P_{k_2}Q_{<k_2}[(1-\phi_{2^{-\frac{i}{2+}}}(u))\psi_2] + (1-\rho_i(r))P_{k_2}Q_{<k_2}[(1-\phi_{2^{-\frac{i}{2+}}}(u))\psi_2],
\end{aligned}$$

where $\rho_i(r)$ localizes smoothly to a disc of radius $\sim 2^{i-10}$ around the origin. Now on account of the fact that $\nabla^{-1}P_{k_2}$ is given by a convolution kernel which decays rapidly outside of a disc of radius $< 2^{\frac{i}{C}}$, we see by means of Proposition 1.1 that

$$||\chi_{i1}(t)\rho_i(r)\nabla^{-1}P_{k_2}Q_{<k_2}[(1-\phi_{2^{-\frac{i}{2+}}}(u))\psi_2]||_{L_x^\infty} \leq C2^{-\frac{3i}{2+}}$$

One then easily concludes that the contribution of this term is negligible: indeed, plugging it into the inner bracket instead of $P_{k_2}\psi_2$ and using schematic notation, we can estimate for example

$$\begin{aligned}
&||\chi_i(t)\nabla_{x,t}P_{k_0}[P_{k_1}Q_{<k_1-100}\delta\psi_1\nabla^{-1}P_kQ_{>\frac{k}{2}}(\chi_{i1}(t)[,])]||_{N[k_0]}\\
&\leq C||\chi_i(t)\nabla_{x,t}P_{k_0}[P_{k_1}Q_{<k_1-100}\delta\psi_1\nabla^{-1}P_kQ_{>\frac{k}{2}}(\chi_{i1}(t)[,])]||_{\dot{X}_{k_0}^{-\frac{1}{2},-1,2}}\\
&\leq C2^{+\frac{k_0}{2}-k}||P_{k_1}Q_{<k_1-100}\delta\psi_1||_{L_t^\infty L_x^2}||P_kQ_{>\frac{k}{2}}(\chi_{i1}(t)[,])||_{L_t^2L_x^2},
\end{aligned}$$

---

[6]More precisely, these errors behave like $[1+2^{-(1-\delta)i}(2^{i-a}-t)]^{-N}$ for $t < 2^{i-a}$ and $[1+2^{-(1-\delta)i}(2^{i+b}-t)]^{-N}$ for $t > 2^{i+b}$, provided $\chi_i(t)$ is supported in $[2^{i-a}, 2^{i+b}]$.

where

$$\begin{aligned}&||P_kQ_{>\frac{k}{2}}(\chi_{i1}(t)[,])||_{L_t^2L_x^2}\\&\leq C||\chi_{i1}(t)(\partial_t+\partial_r)[\rho_i(r)\nabla^{-1}P_{k_2}Q_{<k_2}[(1-\phi_{2^{-\frac{i}{2+}}}(u))\psi_2]]||_{L_t^2L_x^\infty}\\&\qquad\qquad ||(\partial_t-\partial_r)\nabla^{-1}P_{k_3}Q_{<k_3}[(1-\phi_{2^{-\frac{i}{2+}}}(u))\psi_2]||_{L_t^\infty L_x^2}\\&\leq C2^{-(1-\epsilon)i}\end{aligned}$$

Our assumption $|k|<(1+\mu)i$ shows that putting these estimates together gives an acceptable bound. The contribution of $P_{k_1}Q_{\geq k_1-100}\delta\psi_1$ is handled similarly. Now consider the contribution of the term with $\rho_i(r)$ replaced by $(1-\rho_i(r))$. We use the fact (see e. g. [**36**]) that

$$[\frac{x_i}{r}(1-\rho_i(r)),P_{k_2}]" = "2^{-k_2}\nabla(\frac{x_i}{r}(1-\rho_i(r)))$$

where the latter expression stands for a weighted average of translates of the derivatives of $\frac{x_i}{r}(1-\rho_i(r))$. Notice that

$$||\nabla_x(\frac{x_i}{r}(1-\rho_i(r)))||_{L_x^\infty}\leq C2^{-i},$$

hence the contribution of the commutator is treated exactly as the contribution of the term $\rho_i(r)P_{k_2}Q_{<k_2}[(1-\phi_{2^{-\frac{i}{2+}}}(u))\psi_2]$. This finally allows us to move the operator $\partial_t+\partial_r$ past the operator $\nabla^{-1}P_{k_2}Q_{<k_2}$. Arguing as before, one can also move the operator $\chi_{i1}(t)$ past the localizer $\nabla^{-1}P_{k_2}Q_{<k_2}$, generating acceptable error terms. Using lemma 1.7, we can now estimate

$$\begin{aligned}&||P_{k_2}Q_{<k_2}\nabla^{-1}\chi_{i1}(t)(1-\rho_i(r))(\partial_t+\partial_r)[(1-\phi_{2^{-\frac{i}{2+}}}(u))\psi_2]\\&\qquad\qquad(\partial_t-\partial_r)\nabla^{-1}P_{k_3}Q_{<k_3}[(1-\phi_{2^{-\frac{i}{2+}}}(u))\psi_2]||_{L_t^2L_x^2}\\&\leq C||P_{k_2}Q_{<k_2}\nabla^{-1}\chi_{i1}(t)(1-\rho_i(r))(\partial_t+\partial_r)[(1-\phi_{2^{-\frac{i}{2+}}}(u))\psi_2]||_{L_t^2L_x^\infty}\\&\qquad\qquad||(\partial_t-\partial_r)\nabla^{-1}P_{k_3}Q_{<k_3}[(1-\phi_{2^{-\frac{i}{2+}}}(u))\psi_2]||_{L_t^\infty L_x^2}\\&\leq C2^{-\frac{i}{2-}}\end{aligned}$$

Proceeding as before, one deduces from this (and an analogous estimate for the term with $(\partial_t\pm\partial_r)$ interchanged) the following estimate for the full expression:

$$||\chi_i(t)\nabla_{x,t}P_{k_0}[P_{k_1}\delta\psi_1\nabla^{-1}P_kQ_{>\frac{k}{2}}[,]]||_{N[k_0]}\leq C2^{\frac{k_0}{2}-k}2^{-\frac{i}{2-}}\tilde{c}_{k_1}$$

Since $|k|<(1+\mu)i$ and by assumption $k_0\leq k+O(1)$, choosing $\mu$ small enough allows us to get an exponential gain in $i$. This concludes case (I.b).
(I.c): *None of (I.a), (I.b) hold, and* $i\leq C|k_1|$. This then implies $|k_{2,3}|,|k|<<i$, and we shall treat these as $O(1)$. Note that also necessarily $i\leq Ck_1=k_0+O(1)$. One first reduces $P_{k_{2,3}}\psi_{2,3}$ to modulation $<2^{\delta i}$, where $\delta$ is very small but such that $2^{\delta i}>>\max\{|k_{2,3}|,|k|\}$. We shall treat the latter quantities as $O(1)$. To achieve this, one estimates for example

$$\begin{aligned}&||P_{k_0}Q_{<k_0}\nabla_{x,t}\chi_i(t)[P_{k_1}Q_{<\delta i-10}\delta\psi_1\nabla^{-1}(1-I)[P_{k_2}Q_{>\delta i}\psi_2,P_{k_3}Q_{<\delta i-10}\psi_3]]||_{\dot{X}_{k_0}^{-1,-\frac{1}{2},1}}\\&\leq C2^{-\frac{\delta}{2}i}||P_{k_1}\delta\psi_1||_{L_t^\infty L_x^2}||P_{k_2}Q_{>\delta i}R_\nu\psi_2||_{L_t^2L_x^2}||P_{k_3}\psi_3||_{L_t^\infty L_x^\infty}\leq C2^{-\frac{\delta}{2+}i}\tilde{c}_{k_1}\end{aligned}$$

Similarly, one has

$$\begin{aligned}&||P_{k_0}Q_{<k_0}\nabla_{x,t}\chi_i(t)[P_{k_1}Q_{\geq\delta i-10}\delta\psi_1\nabla^{-1}(1-I)[P_{k_2}Q_{>\delta i}\psi_2, P_{k_3}Q_{<\delta i-10}\psi_3]]||_{L_t^1\dot{H}^{-1}}\\&\leq C||P_{k_1}Q_{\geq\delta i-10}\delta\psi_1||_{L_t^2L_x^2}||R_\nu P_{k_2}Q_{>\delta i}\psi_2||_{L_t^2L_x^\infty}||P_{k_3}\psi_3||_{L_t^\infty L_x^\infty}+\text{etc}\\&\leq C2^{-\frac{\delta i}{2+}}\tilde{c}_{k_1}\end{aligned}$$

The estimate when $P_{k_3}Q_{<\delta i-10}\psi_3$ gets replaced by $P_{k_3}Q_{\geq\delta i-10}\psi_3$ is more of the same. Moreover, we have by assumption

$$\begin{aligned}&||P_{k_0}Q_{\geq k_0}\nabla_{x,t}\chi_i(t)[P_{k_1}\delta\psi_1\nabla^{-1}(1-I)[P_{k_2}Q_{>\delta i}\psi_2, P_{k_3}\psi_3]]||_{\dot{X}_{k_0}^{-\frac{1}{2},-1,1}}\\&\leq C2^{-\frac{k_0}{2}}||P_{k_1}\delta\psi_1||_{L_t^ML_x^{2+}}||R_\nu P_{k_2}Q_{>\delta i}\psi_2||_{L_t^\infty L_x^M}||P_{k_3}\psi_3||_{L_t^{2+}L_x^\infty},\end{aligned}$$

which leads to an acceptable estimate. Now we estimate

$$\begin{aligned}&||P_{k_0}Q_{<k_0}\nabla_{x,t}\chi_i(t)[P_{k_1}Q_{<k-10}\delta\psi_1\\&\qquad\nabla^{-1}(1-I)P_k[P_{k_2}Q_{<\delta i}\psi_2, P_{k_3}Q_{<\delta i-10}\psi_3]]||_{\dot{X}_{k_0}^{-1,-\frac{1}{2},1}}\\&\leq C2^{2\delta i}||P_{k_1}\delta\psi_1||_{L_t^\infty L_x^2}||P_{k_2}Q_{<\delta i}\chi_i(t)\psi_2||_{L_t^ML_x^\infty}||P_{k_3}Q_{<\delta i-10}\psi_3||_{L_t^{2+}L_x^\infty}\\&\leq C2^{(2\delta-\frac{1}{2+})i}\tilde{c}_{k_1}\end{aligned}$$

(I.d): *None of (I.a), (I.b), (I.c) hold.* In this case, we may treat all of $k$, $k_i$, $i=1,2,3$ as $O(1)$. The exponential gain in $i$ is again obtained as in the preceding case. This concludes the treatment of case (I).

(II): *The 2nd term.* This term is significantly simpler than the preceding one: note that if at least one of $|k|$,$|k_i|$, $i=0,\dots 3$, is of size at least comparable to $\log_2 T_0$, one gets an exponential gain in $T_0$ from lemma 2.6 in conjunction with lemma 3.1 and the calculations in (I), provided $P_{k_1}\psi$ is of first type. If it is of 2nd type, and $P_{k_3}\psi_3$ of 2nd type as well, one also argues as in (I). If $P_{k_1}\psi_1$ is of 2nd type, but $P_{k_3}\psi_3$ of first type, one uses the estimate

$$||P_k[R_\beta P_{k_2}Q_{<k_2}\delta\psi_2R_jP_{k_3}Q_{<k_3}\psi_3-R_jP_{k_2}Q_{<k_2}\delta\psi_2R_\beta P_{k_3}Q_{<k_3}\psi_3]||_{L_t^\infty L_x^2}\leq C2^k$$

and (with a similar estimate when $Q_{<k_3}$ is replaced by $Q_{\geq k_3}$)

$$\begin{aligned}&||P_k[R_\beta P_{k_2}Q_{\geq k_2}\delta\psi_2R_jP_{k_3}Q_{<k_3}\psi_3-R_jP_{k_2}Q_{\geq k_2}\delta\psi_2R_\beta P_{k_3}Q_{<k_3}\psi_3]||_{L_t^2L_x^2}\\&\leq C2^{\frac{\min\{k,k_2,k_3\}}{2}}2^{-\delta|k_2-k_3|}\frac{\tilde{c}_{k_2}}{\epsilon}\tilde{c}_{k_3}\end{aligned}\tag{3.2}$$

Using the first of these, one gets for example when $k_0=k_1+O(1)$

$$\begin{aligned}&||\nabla_{x,t}P_{k_0}Q_{>k_0}\chi(\frac{t}{T_0})[P_{k_1}\psi_1\triangle^{-1}\sum_{j=1,2}\partial_j(1-I)P_k[R_\beta P_{k_2}Q_{<k_2}\delta\psi_2R_jP_{k_3}Q_{<k_3}\psi_3\\&\qquad-R_jP_{k_2}Q_{<k_2}\delta\psi_2R_\beta P_{k_3}Q_{<k_3}\psi_3]]||_{\dot{X}_{k_0}^{-\frac{1}{2},-1,2}}\\&\leq C2^{-\frac{k_0}{2}}||P_{k_1}\psi_1||_{L_t^2L_x^{2+}}||\triangle^{-1}\sum_{j=1,2}\partial_j(1-I)P_k[R_\beta P_{k_2}Q_{<k_2}\delta\psi_2R_jP_{k_3}Q_{<k_3}\psi_3\\&\qquad-R_jP_{k_2}Q_{<k_2}\delta\psi_2R_\beta P_{k_3}Q_{<k_3}\psi_3]||_{L_t^\infty L_x^M}\end{aligned}$$

which is bounded by

$$2^{(k-k_0)(1-\epsilon)}2^{\delta[\min\{k_2,k_3\}-\max\{k_2,k_3\}]}\tilde{c}_{k_1}\frac{\tilde{c}_{k_2}}{\epsilon}\frac{\tilde{c}_{k_3}}{\epsilon}$$

Assuming $\max\{|k|,|k_i\}$ to be at least comparable to $\delta\log_2 T_0$ and summing over $k, k_{2,3}$ results thus in the estimate $\tilde{c}_{k_1}T_0^{-\mu}$. Using the 2nd of the above inequalities, (3.2), and placing $P_{k_1}\psi_1$ into $L_t^\infty L_x^2$ results in a similar estimate provided $k_0 = k_1 + O(1)$. The remaining frequency interactions $k_0 << k_1$ etc. are handled similarly, as well as the contribution when $Q_{>k_0}$ is replaced by $Q_{\leq k_0}$. Thus assume now that all the occuring frequencies $|k|$, $|k_i|$, $i = 0,\dots,3$ are of size $< \delta\log_2 T_0$. Then we use Proposition 1.1 directly in conjunction with 3.4(b) resp. lemma 3.2, to get

$$||\nabla_{x,t}P_{k_0}Q_{>k_0}\chi(\frac{t}{T_0})[P_{k_1}\psi_1\triangle^{-1}\sum_{j=1,2}\partial_j(1-I)P_k[R_\beta P_{k_2}\delta\psi_2 R_j P_{k_3}\psi_3 - R_j P_{k_2}\delta\psi_2 R_\beta P_{k_3}\psi_3]]||_{\dot{X}_{k_0}^{-\frac{1}{2},-1,2}}$$

$$\leq C||P_{k_1}\chi(\frac{t}{T_0})\psi_1||_{L_t^\infty L_x^\infty}||\triangle^{-1}\sum_{j=1,2}\partial_j(1-I)P_k[R_\beta P_{k_2}\delta\psi_2 R_j P_{k_3}\psi_3 - R_j P_{k_2}\delta\psi_2 R_\beta P_{k_3}\psi_3]||_{L_t^2 L_x^2}$$

$$\leq CT_0^{-\frac{1}{2+}}[\tilde{c}_{k_3}+\tilde{c}_{k_2}]$$

But using the definition of frequency envelope we have $\tilde{c}_{k_{2,3}} \leq CT_0^{\sigma\delta}\tilde{c}_{k_1}$, so we arrive at an acceptable estimate upon summing over the admissible frequency ranges. If we replace $Q_{>k_0}$ by $Q_{\leq k_0}$, we can for example first reduce $P_{k_1}\psi_1$ to modulation $< 2\delta\log_2 T_0$, then reduce both of $P_{k_{2,3}}\psi_{2,3}$ to modulation $< 3\delta\log_2 T_0$, and finally estimate

$$||\nabla_{x,t}P_{k_0}Q_{\leq k_0}\chi(\frac{t}{T_0})[P_{k_1}Q_{k-10<.<2\delta\log_2 T_0}\psi_1$$
$$\triangle^{-1}\sum_{j=1,2}\partial_j(1-I)P_k[R_\beta P_{k_2}Q_{<3\delta\log_2 T_0}\delta\psi_2 R_j P_{k_3}Q_{<3\delta\log_2 T_0}\psi_3 - R_j P_{k_2}Q_{<3\delta\log_2 T_0}\delta\psi_2 R_\beta P_{k_3}Q_{<3\delta\log_2 T_0}\psi_3]]||_{L_t^1\dot{H}^{-1}}$$
$$\leq C||P_{k_1}Q_{k-10<.<2\delta\log_2 T_0}\psi_1||_{L_t^2 L_x^M}||\nabla_{x,t}\nabla^{-1}P_{k_2}Q_{<3\delta\log_2 T_0}\delta\psi_2||_{L_t^M L_x^{2+}}||\nabla_{x,t}\nabla^{-1}P_{k_3}Q_{<3\delta\log_2 T_0}\psi_3||_{L_t^{2+}L_x^\infty}$$

Arranging that $\frac{1}{2}-\frac{1}{2+} >> \delta$ and using the usual properties of the frequency envelope easily results in the desired bound. If one replaces $P_{k_1}Q_{k-10<.<2\delta\log_2 T_0}\psi_1$ by $P_{k_1}Q_{\leq k}\psi_1$, one can estimate the output with respect to $||.||_{\dot{X}^{-1,-\frac{1}{2},1}}$ in the same manner. This finishes case (II) and thereby the large modulation case **(A)**.

**(B): The small modulation case.** We now study the expressions

$$(I):\ \partial^\beta[\delta\psi_\alpha\triangle^{-1}\sum_{j=1,2}\partial_j I[R_\beta\psi_2 R_j\psi_3 - R_j\psi_2 R_\beta\psi_3]]$$

$$(II):\ \partial^\beta[\psi_\alpha\triangle^{-1}\sum_{j=1,2}\partial_j I[R_\beta\delta\psi_2 R_j\psi_3 - R_j\delta\psi_2 R_\beta\psi_3]],$$

as well as the analogous expressions

$$(III):\ \partial^\beta[\delta\psi_\beta\triangle^{-1}\sum_{j=1,2}\partial_j I[R_\beta\psi_2 R_j\psi_3 - R_j\psi_2 R_\beta\psi_3]]$$

$$(IV):\ \partial^\beta[\psi_\beta\triangle^{-1}\sum_{j=1,2}\partial_j I[R_\beta\delta\psi_2 R_j\psi_3 - R_j\delta\psi_2 R_\beta\psi_3]]$$

$$(V) : \partial_\alpha[\delta\psi^\nu \triangle^{-1} \sum_{j=1,2} \partial_j I[R_\nu\psi_2 R_j\psi_3 - R_j\psi_2 R_\nu\psi_3]]$$

$$(VI) : \partial_\alpha[\psi^\nu \triangle^{-1} \sum_{j=1,2} \partial_j I[R_\nu\delta\psi_2 R_j\psi_3 - R_j\delta\psi_2 R_\nu\psi_3]]$$

As is, these terms cannot yet be well estimated, and we need to further decompose the first input into a gradient part and elliptic error term: thus for example we write in term (I)

$$\delta\psi_\alpha = R_\alpha\delta\psi + \delta\chi_\alpha$$

Relegating the error terms involving $\chi_\alpha$ until later, we substitute $R_\alpha\delta\psi$ for $\delta\psi_\alpha$, and similarly for the other terms (II)-(VI). We commence with the sum of first and fifth term in the list:
(I): **(I+V):** As in the large modulation case, we shall have to consider various types of frequency interactions. We also reiterate the decomposition

$$\begin{aligned}
&\chi(\frac{t}{T_0})\partial^\beta[R_\alpha\delta\psi\triangle^{-1}\sum_{j=1,2}\partial_j I[R_\beta\psi_2 R_j\psi_3 - R_j\psi_2 R_\beta\psi_3]] \\
&+\chi(\frac{t}{T_0})\partial_\alpha[R^\nu\delta\psi\triangle^{-1}\sum_{j=1,2}\partial_j I[R_\nu\psi_2 R_j\psi_3 - R_j\psi_2 R_\nu\psi_3]] \\
&= \sum_{i>\log_2 T_0} \chi_i(t)[\partial^\beta[R_\alpha\delta\psi\triangle^{-1}\sum_{j=1,2}\partial_j I[R_\beta\psi_2 R_j\psi_3 - R_j\psi_2 R_\beta\psi_3]] \\
&+\partial_\alpha[R^\nu\delta\psi\triangle^{-1}\sum_{j=1,2}\partial_j I[R_\nu\psi_2 R_j\psi_3 - R_j\psi_2 R_\nu\psi_3]]]
\end{aligned}$$

We frequency-localize this to obtain the following expression:

$$\begin{aligned}
&\chi_i(t)P_{k_0}[\partial^\beta[R_\alpha P_{k_1}\delta\psi_1 P_k\triangle^{-1}\sum_{j=1,2}\partial_j I[R_\beta P_{k_2}\psi_2 R_j P_{k_3}\psi_3 - R_j P_{k_2}\psi_2 R_\beta P_{k_3}\psi_3]] \\
&+\partial_\alpha[R^\nu P_{k_1}\delta\psi_1 P_k\triangle^{-1}\sum_{j=1,2}\partial_j I[R_\nu P_{k_2}\psi_2 R_j P_{k_3}\psi_3 - R_j P_{k_2}\psi_2 R_\nu P_{k_3}\psi_3]]]
\end{aligned}$$

Now we subdivide into the following possibilities:
(I.a): *One of the following options hold:* $i \leq C|k_2|$, $i \leq C|k_3|$, $i \leq C|k_0 - k_1|$, $i \leq C\min\{|k-k_1|, |k-k_2|\}$. In this case, we obtain the desired estimate involving an exponential gain in $i$ from 3.4(c) as well as lemma 2.6 if both $P_{k_{2,3}}\psi_{2,3}$ are of the first type. If at least one of them is of the 2nd type, this is again straightforward due to the strong estimates satisfied by these: then we have

$$\begin{aligned}
&||P_k\triangle^{-1}\sum_{j=1,2}\partial_j I[R_\beta P_{k_2}\psi_2 R_j P_{k_3}\psi_3 - R_j P_{k_2}\psi_2 R_\beta P_{k_3}\psi_3]||_{L_t^1 L_x^\infty} \\
&\leq C2^{\delta[\min\{k,k_2,k_3\}-\max\{k,k_2,k_3\}]}\frac{\tilde{c}_{k_2}}{\epsilon}\frac{\tilde{c}_{k_3}}{\epsilon}
\end{aligned}$$

$$\begin{aligned}
&||P_k\triangle^{-1}\sum_{j=1,2}\partial_j I[R_\beta P_{k_2}\psi_2 R_j P_{k_3}\psi_3 - R_j P_{k_2}\psi_2 R_\beta P_{k_3}\psi_3]||_{L_t^1 L_x^{2+}} \\
&\leq C2^{-k(1-\epsilon)}2^{\delta[\min\{k,k_2,k_3\}-\max\{k,k_2,k_3\}]}\frac{\tilde{c}_{k_2}}{\epsilon}\frac{\tilde{c}_{k_3}}{\epsilon}
\end{aligned}$$

From this we get

$$
\begin{aligned}
&||P_{k_0}Q_{<k_0}[\partial^\beta[R_\alpha P_{k_1}\delta\psi_1 \\
&\qquad P_k\triangle^{-1}\sum_{j=1,2}\partial_j I[R_\beta P_{k_2}\psi_2 R_j P_{k_3}\psi_3 - R_j P_{k_2}\psi_2 R_\beta P_{k_3}\psi_3]]||_{L_t^1\dot{H}^{-1}} \\
&\qquad\qquad \leq C2^{\delta[\min\{k,k_0,\dots,k_3\}-\max\{k,k_0,\dots,k_3\}]}\frac{1}{\epsilon^2}\prod_{i=1,2,3}\tilde{c}_{k_i}
\end{aligned}
$$

The contribution when one has $P_{k_0}Q_{\geq k_0}$ in front is even simpler and left out. Term (V) is treated by exact analogy.
(I.b): *$i \leq C|k|$, $i \leq C|k_1|$, and none of the properties in (I.a) hold.* This implies $k, k_1 \leq C - i$. We may and shall assume $k - k_1 = O(1)$, $k_0 - k_1 = O(1)$. Also, we may and shall assume that $|k_{2,3}| = O(1)$. We shall again treat term (I), term (V) being treated analogously. We start out by observing that we may assume $k > -i(1+\mu)$ for any $\mu > 0$. Indeed, assume the opposite. Again considering term (I), we have

$$
\begin{aligned}
&||P_{k_0}Q_{<k_0}\chi_i(t)[\partial^\beta Q_{>k_0+100}[R_\alpha P_{k_1}\delta\psi_1 P_k \\
&\qquad \triangle^{-1}\sum_{j=1,2}\partial_j I[R_\beta P_{k_2}\psi_2 R_j P_{k_3}\psi_3 - R_j P_{k_2}\psi_2 R_\beta P_{k_3}\psi_3]]||_{L_t^1\dot{H}^{-1}} \\
&\leq C\sum_{k_0+100\leq j\leq -i+O(1)} ||P_{k_0}Q_{<k_0}\chi_i(t)[\partial^\beta Q_j[R_\alpha P_{k_1}Q_{j+O(1)}\delta\psi_1 P_k \\
&\qquad \triangle^{-1}\sum_{j=1,2}\partial_j I[R_\beta P_{k_2}\psi_2 R_j P_{k_3}\psi_3 - R_j P_{k_2}\psi_2 R_\beta P_{k_3}\psi_3]]||_{L_t^1\dot{H}^{-1}} \\
&\leq C\sum_{k_0+100\leq j\leq -i+O(1)} 2^j||\chi_i(t)||_{L_t^2}||R_\alpha P_{k_1}Q_{j+O(1)}\delta\psi_1||_{L_t^2L_x^2} \\
&\qquad\qquad ||P_k[\nabla^{-1}P_{k_2}\psi_2 P_{k_3}\psi_3]||_{L_t^\infty L_x^\infty}
\end{aligned}
$$

One checks that this is estimated by $\leq C\tilde{c}_{k_1}2^{k_0}$, which is acceptable. Similarly, we estimate

$$
\begin{aligned}
&||P_{k_0}Q_{<k_0}\chi_i(t)[\partial^\beta Q_{\leq k_0+100}[R_\alpha P_{k_1}\delta\psi_1 \\
&\qquad P_k\triangle^{-1}\sum_{j=1,2}\partial_j I[R_\beta P_{k_2}\psi_2 R_j P_{k_3}\psi_3 - R_j P_{k_2}\psi_2 R_\beta P_{k_3}\psi_3]]||_{L_t^1\dot{H}^{-1}} \\
&\leq C||\chi_i(t)||_{L_t^2}||P_{k_1}\delta\psi_1||_{L_t^\infty L_x^\infty} \\
&\qquad ||P_k\triangle^{-1}\sum_{j=1,2}\partial_j I[R_\beta P_{k_2}\psi_2 R_j P_{k_3}\psi_3 - R_j P_{k_2}\psi_2 R_\beta P_{k_3}\psi_3]||_{L_t^2L_x^2},
\end{aligned}
$$

which, upon using 3.4(b) as well as lemma 3.2, can be estimated by $\leq C\tilde{c}_{k_0}2^{\frac{k_0+i}{2}}$, which is acceptable.
The estimates when $Q_{<k_0}$ is replaced by $Q_{\geq k_0}$ are similar. Thus we now assume that $|k| < i(1+\mu)$. Arguing as in case **(A)**(I.b), we may replace $P_k I$ be $P_k Q_{<\frac{k}{2}}$ while only generating acceptable error terms. We may then move the multiplier $\chi_i(t)$ past the operator $P_k\nabla^{-1}Q_{<\frac{k}{2}}$, and transform the latter back into $\nabla^{-1}P_k I$ innocuously. Proceeding as in [**23**], we intend to exploit the null-structure of the expression. Before being able to do so, we need to effect a few more reductions: we need to reduce the modulation of the output and first input to size $< 2^{k_{0,1}}$,

respectively. For this, note that

$$
\begin{aligned}
&||P_{k_0}Q_{\geq k_0+100}[\partial^\beta[R_\alpha P_{k_1}\delta\psi_1 P_k \\
&\qquad \triangle^{-1}\sum_{j=1,2}\partial_j I[R_\beta P_{k_2}\chi_i(t)\psi_2 R_j P_{k_3}\psi_3 - R_j P_{k_2}\chi_i(t)\psi_2 R_\beta P_{k_3}\psi_3]]||_{\dot{X}_{k_0}^{-\frac{1}{2},-1,2}} \\
&\leq C||P_{k_0}Q_{\geq k_0+100}[\partial^\beta[R_\alpha P_{k_1}Q_{\geq k_1}\delta\psi_1 P_k \\
&\qquad \triangle^{-1}\sum_{j=1,2}\partial_j I[R_\beta P_{k_2}\chi_i(t)\psi_2 R_j P_{k_3}\psi_3 - R_j P_{k_2}\chi_i(t)\psi_2 R_\beta P_{k_3}\psi_3]]||_{\dot{X}_{k_0}^{-\frac{1}{2},-1,2}} \\
&\leq C2^{-\frac{k_0}{2}}||R_\alpha P_{k_1}Q_{\geq k_1}\delta\psi_1||_{L_t^2 L_x^\infty}||P_k[\nabla^{-1}P_{k_2}\psi_2 P_{k_3}R_\beta\psi_3]||_{L_t^\infty L_x^2} \\
&\leq C2^k\tilde{c}_{k_1}[\frac{\tilde{c}_{k_2}}{\epsilon}+\frac{\tilde{c}_{k_3}}{\epsilon}].
\end{aligned}
$$

We can also reduce $P_{k_{2,3}}\psi_{2,3}$ to modulation $< 2^{\max\{k_2,k_3\}+O(1)}$. For this, note that

$$
\begin{aligned}
&||P_k I[\nabla^{-1}P_{k_2}\psi_2 P_{k_3}Q_{>\max\{k_2,k_3\}+100}\psi_3]||_{L_t^1 L_x^2} \\
&\leq C\sum_{j>\max\{k_2,k_3\}+100} 2^k||\nabla^{-1}P_{k_2}Q_{j+O(1)}\psi_2||_{L_t^2 L_x^2}||P_{k_3}Q_j\psi_3||_{L_t^2 L_x^2} \\
&\leq C\sum_{j>\max\{k_2,k_3\}+100} 2^k 2^{-(2-2\mu)j}2^{(1-2\mu)\max\{k_2,k_3\}} \\
&\qquad\qquad ||\nabla^{-1}P_{k_2}\psi_2||_{\dot{X}_{k_2}^{-(\frac{1}{2}-\mu),1-\mu,1}}||P_{k_3}\psi_3||_{\dot{X}_{k_3}^{-(\frac{1}{2}-\mu),1-\mu,1}} \\
&\leq C2^{-k_2}\frac{\tilde{c}_{k_2}}{\epsilon}\frac{\tilde{c}_{k_3}}{\epsilon}
\end{aligned}
$$

From this one deduces that for $a=\max\{k_2,k_3\}+100$

$$
\begin{aligned}
&||P_{k_0}Q_{<k_0+O(1)}[\partial^\beta[R_\alpha P_{k_1}\delta\psi_1 \\
&P_k\triangle^{-1}\sum_{j=1,2}\partial_j I[R_\beta P_{k_2}\chi_i(t)\psi_2 R_j P_{k_3}Q_{>a}\psi_3 - R_j P_{k_2}\chi_i(t)\psi_2 R_\beta P_{k_3}Q_{>a}\psi_3]]||_{L_t^1\dot{H}^{-1}} \\
&\leq C||R_\alpha P_{k_1}Q_{<k_1+O(1)}\delta\psi_1||_{L_t^\infty L_x^\infty}||P_k I[\nabla^{-1}P_{k_2}\psi_2 P_{k_3}Q_{>\max\{k_2,k_3\}+100}\psi_3]||_{L_t^1 L_x^2} \\
&\leq C2^{k-k_2}\tilde{c}_{k_1}\frac{\tilde{c}_{k_2}}{\epsilon}\frac{\tilde{c}_{k_3}}{\epsilon}
\end{aligned}
$$

One can sum over the appropriate range of $k_{2,3}$, deducing the desired estimate. We shall always assume these reductions of modulation, but sometimes omit them to simplify notation. Now we expand the null-structure as in [**22**], [**23**]: schematically we have

$$
\begin{aligned}
&2\sum_{j=1}^{2}\triangle^{-1}\partial_j[R_\nu f R_j g - R_j f R_\nu g]\partial^\nu h \\
&=\sum_{j=1}^{2}\Box[\triangle^{-1}\partial_j[\nabla^{-1}f R_j g]h] - \sum_{j=1}^{2}\Box\triangle^{-1}\partial_j[\nabla^{-1}f R_j g]h \\
&-\sum_{j=1}^{2}\triangle^{-1}\partial_j[\nabla^{-1}f R_j g]\Box h - \nabla^{-1}f\Box((\nabla^{-1}g)h) \\
&+\nabla^{-1}f\Box(\nabla^{-1}g)h + \nabla^{-1}f(\nabla^{-1}g)\Box h,
\end{aligned}
\tag{3.3}
$$

$$(3.4)\quad \begin{aligned}&\sum_{j=1}^{2}\triangle^{-1}\partial_j\partial^\nu[R_\nu fR_jg-R_jfR_\nu g]h\\&=\sum_{j=1}^{2}[\triangle^{-1}\partial_j\Box[\nabla^{-1}fg]h-\frac{1}{2}\Box[\nabla^{-1}f\nabla^{-1}g]h\\&+\frac{1}{2}\Box\nabla^{-1}f\nabla^{-1}gh-\frac{1}{2}\nabla^{-1}f\Box gh]\end{aligned}$$

The first of these identities is useful when the outer derivative $\partial^\beta$ falls on the first input $R_\alpha\delta\psi_1$. The 2nd is useful provided the outer derivative lands on the inner square bracket. We shall treat each of these terms. Clearly, the terms in the 2nd expansion are almost identical to the ones in the first. We treat the first in detail, the 2nd being treated similarly.

(I.b.1) *The first term in the expansion.* This is the expression

$$P_{k_0}Q_{<k_0}\Box[R_\alpha P_{k_1}Q_{<k_0}\delta\psi_1P_k\triangle^{-1}\sum_{j=1,2}\partial_jI[\nabla^{-1}P_{k_2}\chi_i(t)\psi_2R_jP_{k_3}\psi_3]]$$

This is straightforward to estimate: we have

$$\begin{aligned}&||P_{k_0}Q_{<k_0}\Box[R_\alpha P_{k_1}Q_{<k_0}\delta\psi_1P_k\triangle^{-1}\sum_{j=1,2}\partial_jI[\nabla^{-1}P_{k_2}\chi_i(t)\psi_2R_jP_{k_3}\psi_3]]||_{\dot{X}_{k_0}^{-1-\frac{1}{2},1}}\\&\leq C2^{-\frac{k_0}{2}}||R_\alpha P_{k_1}Q_{<k_0}\delta\psi_1||_{L_t^\infty L_x^M}||\nabla^{-1}P_{k_2}\chi_i(t)\psi_2||_{L_t^{2+}L_x^\infty}||P_{k_3}\psi_3||_{L_t^ML_x^{2+}}\leq C2^{\frac{k_1}{2+}}\tilde{c}_{k_1}\end{aligned}$$

(I.b.2) *The 2nd term in the expansion.* This is the expression

$$P_{k_0}Q_{<k_0}[R_\alpha P_{k_1}Q_{<k_0}\delta\psi_1P_k\triangle^{-1}\sum_{j=1,2}\partial_j\Box I[\nabla^{-1}P_{k_2}\chi_i(t)\psi_2R_jP_{k_3}\psi_3]]$$

This turns out to be significantly more complicated. The reason for this is that we need to exploit the bilinear inequality 3.4(g); using Strichartz type norms here appears to result in a loss in the low frequencies, or in $i$. The only way we can possibly squeeze out a small gain in $i$ is to exploit the temporal cutoff $\chi_i(t)$ applied to $P_{k_2}\psi_2$. This is a non-trivial task on account of the fact that the only way to place the inner bracket [,] into $L_t^2L_x^2$ appears to involve null-frame spaces. Our main tool for this is the following lemma:

LEMMA 3.3. *Let $C$ be a sufficiently large number. The following limits hold: for any $k$ with $|k|<\frac{\epsilon}{1000C}i$, and arbitrary $\epsilon>0$,*

$$\lim_{i\to\infty}||P_kQ_{[k-(1-\epsilon)i-C,k]}[\chi_i(t)\psi_\nu]||_{\dot{X}_k^{0,\frac{1}{2},\infty}}=0$$

*More precisely, for appropriate $\mu(\epsilon)>0$, we have*

$$||P_kQ_{[k-(1-\epsilon)i-C,k]}[\chi_i(t)\psi_\nu]||_{\dot{X}_k^{0,\frac{1}{2},\infty}}\leq C2^{-\mu(\epsilon)i}$$

*Next, denote by $\chi_{i,\kappa}^c$ a smooth bump function which localizes to the complement in the $2^{\epsilon i}/C$-neighborhood of the (physical) light cone of a slab of length $2^{i-1}$ centered at time $t=2^i$, angular opening $2\kappa$ with $|\kappa|\sim 2^{i(\frac{\epsilon-1}{2})}$ and distance $\leq C2^{\epsilon i}/C$ from the light cone. Then with the same assumption on $k$,*

$$\lim_{i\to\infty}\sum_{\pm}\sup_{l\in[\frac{\epsilon-1}{2}i,-10]}2^{-\frac{l}{2}}(\sum_{\tilde{\kappa}\in K_l}||P_{k,\tilde{\kappa}}\sum_{\kappa\in K_{\frac{\epsilon-1}{2}i}}\chi_{i,\mp\kappa}^c(t,x)$$
$$P_{k,\kappa}Q_{<k+i(\epsilon-1)-C}^{\pm}[\chi_i(t)\psi_\nu]||_{PW[\pm\tilde{\kappa}]}^2)^{\frac{1}{2}}=0$$

*More precisely, this quantity decays like $2^{-\mu i}$ for suitable $\mu > 0$.*

PROOF. : We first estimate

$$||\nabla_{x,t}P_kQ_{>k-(1-\epsilon)i}[\chi_i(t)\frac{\mathbf{x}}{\mathbf{y}}]||_{A[k]},\ ||\nabla_{x,t}P_kQ_{>k-(1-\epsilon)i}[\chi_i(t)\ln \mathbf{y}]||_{A[k]}$$

for arbitrary $k \in \mathbf{Z}$ with $|k| < \frac{\epsilon}{C}i$. Observe that

$$\Box[\chi_i(t)\frac{\mathbf{x}}{\mathbf{y}}] = \chi_i''(t)\frac{\mathbf{x}}{\mathbf{y}} + \chi_i'(t)\partial_t(\frac{\mathbf{x}}{\mathbf{y}}) + \chi_i(t)\Box[\frac{\mathbf{x}}{\mathbf{y}}]$$

Therefore, we obtain for $j \in [k-(1-\epsilon)i, k]$:

$$||P_kQ_j\nabla_x\chi_i(t)\frac{\mathbf{x}}{\mathbf{y}}||_{A[k]} \leq C2^{-\frac{j}{2}}[||P_k(\chi_i''(t)\frac{\mathbf{x}}{\mathbf{y}})||_{L_t^2L_x^2} + ||P_k(\chi_i'(t)\partial_t(\frac{\mathbf{x}}{\mathbf{y}}))||_{L_t^2L_x^2} + ||P_kQ_j(\chi_i(t)\Box[\frac{\mathbf{x}}{\mathbf{y}}])||_{L_t^2L_x^2}]$$

The first two terms on the right hand side are elementary to estimate, using Holder's inequality, finite propagation speed and the energy inequality:

$$||P_k(\chi_i''(t)\frac{\mathbf{x}}{\mathbf{y}})||_{L_t^2L_x^2} \leq C||\chi_i''(t)||_{L_t^2}||P_k(\frac{\mathbf{x}}{\mathbf{y}})||_{L_t^\infty L_x^2} \leq C2^{\frac{i}{2}}2^{-2i}2^i \leq C2^{-\frac{i}{2}}$$

$$||P_k(\chi_i'(t)\partial_t(\frac{\mathbf{x}}{\mathbf{y}}))||_{L_t^2L_x^2} \leq C||\chi_i'(t)||_{L_t^2}||\partial_t(\frac{\mathbf{x}}{\mathbf{y}})||_{L_t^\infty L_x^2}. \leq C2^{-\frac{i}{2}}$$

We conclude that the contribution from these terms is at most

$$\leq C\sum_{j>k-i(1-\epsilon)} 2^{-\frac{j}{2}}2^{-\frac{i}{2}} \leq C2^{-\frac{\epsilon}{2+}i}.$$

Proceeding to the last term above, we use the null-structure in (1.2) as well as Proposition 1.1 to get

$$||P_k(\chi_i(t)\Box(\frac{\mathbf{x}}{\mathbf{y}})||_{L_t^2L_x^2}$$
$$\leq C||\chi_i(t)\frac{(\partial_t+\partial_r)\mathbf{x}}{\mathbf{y}}||_{L_t^2L_x^\infty}||\chi_i(t)\frac{(\partial_t-\partial_r)\mathbf{x}}{\mathbf{y}}||_{L_t^\infty L_x^2} + \text{etc} \leq C2^{-i}$$

This establishes the claim for $\frac{\mathbf{x}}{\mathbf{y}}$ because of 3.4(d). As far as $\partial_tP_kQ_{>k+(\epsilon-1)i}(\frac{\mathbf{x}}{\mathbf{y}})$ is concerned this only differs from the preceding as far as the estimate for $P_kQ_{\geq k}R_0\partial_t(\frac{\mathbf{x}}{\mathbf{y}})$. This is treated by using the equation for $\Box(\frac{\mathbf{x}}{\mathbf{y}})$ and arguing as before. The estimate for $\ln \mathbf{y}$ is similar. We now establish the 2nd inequality stated in the lemma provided $\psi_\nu$ is replaced by $\nabla_{x,t}\frac{\mathbf{x}}{\mathbf{y}}$, $\nabla_{x,t}\ln \mathbf{y}$. First, let $\phi \in C^\infty(\mathbf{R}^{2+1})$ be a rotationally symmetric free wave with $\phi[0,x] = (0, g(x))$. We use the representation formula

$$P_{k,\kappa}\phi(t,x) = c\int_{S^1}\mathcal{F}(\hat{g}(|\xi|)m_k(|\xi|)|\xi|^2)a_\kappa(\omega)(-t+x\cdot\omega)d\omega - c\int_{S^1}\mathcal{F}(\hat{g}(|\xi|)m_k(|\xi|)|\xi|^2)a_\kappa(\omega)(t+x\cdot\omega)d\omega$$

where(committing abuse of notation) we wrote $\hat{g}(\xi) = \hat{g}(|\xi|)$. Now assume that $||(1+\nabla_{|\xi|}^\alpha)\hat{g}(|\xi|)||_{L^2_{|\xi|}} < C$. This implies that

$$||\chi_i(.)\mathcal{F}(\hat{g}(|\xi|))(.)||_{L^2} \leq C2^{-\alpha i}$$

Now observe that on the support of $\chi_{i,\mp\kappa}^c$, we have

$$|t \pm x\cdot\omega| \geq -|t-|x|| + ||x| \pm x\cdot\omega| \geq ||t-|x|| - |x|\sin < x, \mp\omega >^2| \geq c2^{i\epsilon}$$

Next, observe that the multiplier $\chi^c_{i,\mp\kappa}(t,x)$ smears out the Fourier support in the angular direction by an amount $\sim 2^{-\frac{(1+\epsilon)i}{2}}$, while $|\kappa|\sim 2^{\frac{i(\epsilon-1)}{2}}$ and $|k|<\frac{\epsilon}{C}i$. This entails that one can include an operator $P'_{k,\kappa}$ with the same properties as $P_{k,\kappa}$ in the expression and reason as follows[7]:

$$2^{-\frac{l}{2}}||P_{k,\tilde\kappa}\sum_{\kappa\in K_{\frac{(\epsilon-1)}{2}i}}\chi^c_{i,\mp\kappa}(t,x)P_{k,\kappa}Q^{\pm}\phi||_{PW[\pm\tilde\kappa]}$$

$$=2^{-\frac{l}{2}}||P_{k,\tilde\kappa}\sum_{\kappa\in K_{\frac{(\epsilon-1)}{2}i}}P'_{k,\kappa}\chi^c_{i,\mp\kappa}(t,x)P_{k,\kappa}Q^{\pm}\phi||_{PW[\pm\tilde\kappa]}$$

$$\leq C2^{i\frac{1-\epsilon}{4}}(\sum_{\kappa\in K_{\frac{(\epsilon-1)i}{2}},\kappa\cap 2\tilde\kappa\neq\emptyset}||\chi^c_{i,\mp\kappa}(t,x)P_{k,\kappa}Q^{\pm}\phi||^2_{PW[\pm\kappa]})^{\frac12}$$

Using Plancherel's theorem, we can further estimate this as

$$\leq C2^{i\frac{1-\epsilon}{4}}(\sum_{\kappa\in K_{\frac{(\epsilon-1)i}{2}},\kappa\cap 2\tilde\kappa\neq\emptyset}[\int_{\omega\in\kappa}||\chi_{t\pm x\cdot\omega>2^{\epsilon i}}\mathcal{F}(\hat g(|\xi|))(t\pm x\cdot\omega)||_{L^2}a_\kappa(\omega)d\omega]^2)^{\frac12}$$

$$\leq C2^{-\epsilon\alpha i}(\sum_{\kappa\in K_{\frac{(\epsilon-1)i}{2}},\kappa\cap 2\tilde\kappa\neq\emptyset}[\int_{\omega\in\kappa}||(1+\nabla^\alpha_{|\xi|})\hat g(|\xi|)||_{L^2_{|\xi|}}a_\kappa(\omega)d\omega]^2)^{\frac12}$$

Using Cauchy Schwarz' inequality, this leads to the estimate

$$2^{-\frac{l}{2}}\sum_{\pm}(\sum_{\tilde\kappa\in K_l}||P_{k,\tilde\kappa}\sum_{\kappa\in K_{\frac{(\epsilon-1)}{2}i}}\chi^c_{i,\mp\kappa}(t,x)P_{k,\kappa}Q^{\pm}\phi||^2_{PW[\pm\tilde\kappa]})^{\frac12}$$

$$\leq C2^{-\alpha\epsilon i}||(1+\nabla^\alpha_{|\xi|})\hat g(|\xi|)||_{L^2_{|\xi|}}$$

Now we proceed to the inhomogeneous situation at hand: let $S(t)$ denote the free wave propagator, i. e. $\Box[S(t)(f,g)]=0$, $S(0)(f,g)=f$, $\partial_t[S(0)(f,g)]=g$. Also, let $U(t)g=S(t)(0,f)$. Then we can write

$$\frac{\mathbf{x}}{\mathbf{y}}(t,.)=S(t)(\frac{\mathbf{x}}{\mathbf{y}}(0,.),\partial_t[\frac{\mathbf{x}}{\mathbf{y}}](0,.))+\int_0^t U(t-s)\Box[\frac{\mathbf{x}}{\mathbf{y}}](s,.)ds$$

Reasoning as above, we immediately get the desired estimate for the linear part. As concerns the inhomogeneity, we decompose the integral as

$$\int_0^t U(t-s)\Box[\frac{\mathbf{x}}{\mathbf{y}}](s,.)ds=\int_0^{2^{\frac{\epsilon}{C}i}}U(t-s)\Box[\frac{\mathbf{x}}{\mathbf{y}}](s,.)ds+\int_{2^{\frac{\epsilon}{C}i}}^t U(t-s)\Box[\frac{\mathbf{x}}{\mathbf{y}}](s,.)ds$$

Then we observe that from the argument given above we have for $l\in[\frac{(\epsilon-1)i}{2},-10]$

$$2^{-\frac{l}{2}}(\sum_{\tilde\kappa\in K_l}||P_{k,\tilde\kappa}\sum_{\kappa\in K_{\frac{\epsilon-1}{2}i}}\chi^c_{i,\mp\kappa}P_{k,\kappa}Q^{\pm}[\int_0^{2^{\frac{\epsilon}{C}i}}U(t-s)\Box[\frac{\mathbf{x}}{\mathbf{y}}](s,.)ds]||^2_{PW[\pm\tilde\kappa]})^{\frac12}$$

$$\leq C2^{-\epsilon\alpha i}2^{-k}||x^\alpha\Box[\frac{\mathbf{x}}{\mathbf{y}}]||_{L^1_tL^2_x}\leq C2^{-\frac{\epsilon}{2}\alpha i},$$

---

[7]Let $Q^{\pm}$ microlocalize to the upper or lower half-space $\tau><0$.

provided $\frac{1}{C} << \alpha$ and we choose $\alpha > 0$ small enough, by the proof of Corollary 1.1 and finite propagation speed. Next, we can estimate by using 3.4(d):

$$2^{\frac{1-\epsilon}{4}i}\Big(\sum_{\kappa\in K_{\frac{\epsilon-1}{2}i}}||P_{k,\pm\kappa}Q^{\pm}_{<k+(\epsilon-1)i}\chi_i(t)[\int_{2^{\frac{\epsilon}{C}i}}^{t}U(t-s)\Box[\frac{\mathbf{x}}{\mathbf{y}}](s,.)ds]||^2_{PW[\tilde{\kappa}]}\Big)^{\frac{1}{2}}$$

$$\leq C||\chi_{t>2^{\frac{\epsilon}{C}i}}\Box[\frac{\mathbf{x}}{\mathbf{y}}]||_{L_t^1L_x^2}\leq C2^{-\frac{\epsilon}{C}\alpha i}$$

Call the integral in the preceding $\psi$. From this we get control over the 2nd more complicated norm in the lemma: using Cauchy-Schwartz and the fact that $||.||_{PW[\kappa]}$ is essentially unaffected by multiplication with bounded functions, we get

$$\begin{aligned}(3.5)\qquad &2^{-\frac{l}{2}}\Big(\sum_{\tilde{\kappa}\in K_l}||P_{k,\tilde{\kappa}}\sum_{\kappa\in K_{\frac{\epsilon-1}{2}i}}\chi^c_{i,\mp\kappa}P_{k,\kappa}Q^{\pm}_{<k+(\epsilon-1)i}\psi||^2_{PW[\pm\tilde{\kappa}]}\Big)^{\frac{1}{2}}\\ &\leq C2^{\frac{1-\epsilon}{4}i}\Big(\sum_{\tilde{\kappa}\in K_l}\sum_{\kappa\in K_{\frac{\epsilon-1}{2}i},\kappa\subset 2\tilde{\kappa}}||P_{k,\kappa}Q^{\pm}_{<k+(\epsilon-1)i}\psi||^2_{PW[\pm\kappa]}\Big)^{\frac{1}{2}},\end{aligned}$$

and we just bounded this expression. The estimate for $\ln\mathbf{y}$ is of course analogous. Let $\mu_1(\epsilon) = \frac{\epsilon}{C}\alpha$. Now we need to transfer these statements to $\psi_\nu$. We recall the identity

$$\psi_\nu = \Big(\frac{\partial_\nu\mathbf{x}}{\mathbf{y}} + i\frac{\partial_\nu\mathbf{y}}{\mathbf{y}}\Big)e^{i\sum_{j=1,2}\triangle^{-1}\partial_j(\frac{\partial_j\mathbf{x}}{\mathbf{y}})}$$

We observe as usual that $\frac{\partial_\nu\mathbf{x}}{\mathbf{y}} = \partial_\nu(\frac{\mathbf{x}}{\mathbf{y}}) - \frac{\mathbf{x}}{\mathbf{y}}\frac{\partial_\nu\mathbf{y}}{\mathbf{y}}$. For example consider the term $\frac{\mathbf{x}}{\mathbf{y}}\frac{\partial_\nu\mathbf{y}}{\mathbf{y}}e^{i\sum_{j=1,2}\triangle^{-1}\partial_j(\frac{\partial_j\mathbf{x}}{\mathbf{y}})}$, the other terms being treated similarly. One expands the exponential in a Taylor series, which results in schematic terms of the form

$$a_k\psi\nabla^{-1}\psi_1\nabla^{-1}\psi_2\ldots\nabla^{-1}\psi_k,$$

where $\nabla^{-1}\psi$ stands for expressions like $\frac{\mathbf{x}}{\mathbf{y}}$, $\nabla^{-1}(\frac{\mathbf{x}}{\mathbf{y}}\frac{\partial_i\mathbf{y}}{\mathbf{y}})$, and $\psi$ stands for either $\frac{\partial_\nu\mathbf{y}}{\mathbf{y}}$ or $\frac{\partial_\nu\mathbf{x}}{\mathbf{y}}$. Also note that the coefficients of these expressions decay faster than exponentially. Now apply a localizer $P_{k_0}$, $|k_0| < \frac{\epsilon i}{1000C}$ in front. We redefine $C$ so large that $\frac{\epsilon}{C} << \mu_1(\epsilon)$, the latter coming from the preceding computation. We claim that if one of the input frequencies has absolute value greater than $i\delta$ for suitable $\delta > \frac{\epsilon}{500C}$, one obtains an exponential gain in $i$ for the norms in the statement of the lemma. This is done inductively: write the expression under consideration as

$$P_k[P_{k_1}(\psi_1\nabla^{-1}\psi_2\ldots\nabla^{-1}\psi_{k-1})\nabla^{-1}P_{k_2}\psi_k]$$

If $k_1 > \frac{\epsilon i}{1000C} + 5$, so is $k_2$. If $P_{k_1}(...)$, $P_{k_2}\psi_2$ are both of first type, one estimates using 3.4(a)

$$||P_k[P_{k_1}(\psi_1\nabla^{-1}\psi_2\ldots\nabla^{-1}\psi_{k-1})\nabla^{-1}P_{k_2}\psi_k]||_{\dot{X}_k^{0,\frac{1}{2},1}}$$

$$\leq C||P_{k_1}(\psi_1\nabla^{-1}\psi_2\ldots\nabla^{-1}\psi_{k-1})||_{A[k_1]}||P_{k_2}\psi_2||_{A[k_2]}\leq C\frac{\tilde{c}_{k_1}}{\epsilon}\frac{\tilde{c}_{k_2}}{\epsilon}$$

If one of them is of 2nd type, one places this into $L_t^2L_x^{2+}$ to control the portion of the output at modulation $< 2^{k+100}$ (the other portion being controlled by theorem 2.3.). One obtains the same bound, and our decay assumptions on the frequency envelope yield the claim, provided one shows that the 2nd more complicated norm in the statement of the lemma is controlled by $||.||_{\dot{X}_k^{0,\frac{1}{2},1}}$. This follows from (2.3) as well as the preceding computation (3.5). In case $k_2 < -\frac{\epsilon i}{1000C} - 5$, first assume $P_{k_2}\psi_2$ to be of 2nd type. If $k_2 \geq j - 10$, we estimate

$$||P_k Q_j[P_{k_1}\psi_1 \nabla^{-1} P_{k_2}\psi_2]||_{\dot{X}_k^{0,\frac{1}{2},\infty}} \leq C2^{\frac{j}{2}}||P_{k_1}\psi_1||_{L_t^\infty L_x^2}||\nabla^{-1}P_{k_2}\psi_2||_{L_t^2 L_x^\infty} \leq C2^{\frac{j-k_2}{2}}\tilde{c}_{k_2}$$

If $k_2 < j - 10$, we estimate

$$||P_k Q_j[P_{k_1}\psi_1 \nabla^{-1} P_{k_2}\psi_2]||_{\dot{X}_k^{0,\frac{1}{2},\infty}} \leq C||P_k Q_j[P_{k_1}\psi_1 \nabla^{-1} P_{k_2} Q_{\geq j-10}\psi_2]||_{\dot{X}_k^{0,\frac{1}{2},\infty}}$$
$$+ ||P_k Q_j[P_{k_1} Q_{\geq j-10}\psi_1 \nabla^{-1} P_{k_2} Q_{<j-10}\psi_2]||_{\dot{X}_k^{0,\frac{1}{2},\infty}}$$

Then we have

$$||P_k Q_j[P_{k_1}\psi_1 \nabla^{-1} P_{k_2} Q_{\geq j-10}\psi_2]||_{\dot{X}_k^{0,\frac{1}{2},\infty}} \leq C||P_{k_1}\psi_1||_{L_t^\infty L_x^2}||\nabla^{-1}P_{k_2}Q_{\geq j-10}\psi_2||_{L_t^2 L_x^\infty}$$
$$\leq C2^{\frac{1}{2+}(k_2-j)}\tilde{c}_{k_2}$$

Also, we have

$$||P_k Q_j[P_{k_1} Q_{\geq j-10}\psi_1 \nabla^{-1} P_{k_2} Q_{<j-10}\psi_2]||_{\dot{X}_k^{0,\frac{1}{2},\infty}}$$
$$\leq C2^{\frac{j}{2}}||P_{k_1}Q_{\geq j-10}\psi_1||_{L_t^2 L_x^2}||\nabla^{-1}P_{k_2}Q_{<j-10}\psi_2||_{L_t^\infty L_x^\infty} \leq C\tilde{c}_{k_2}$$

Control over the more complicated norm in the lemma is a bit more difficult, and follows from a computation performed in the appendix, in addition to the calculation (3.5) performed above: one again obtains the bound $\leq C\tilde{c}_{k_2}$, and the claim follows by summing over $k_2 < -\frac{\epsilon i}{1000C} - 5$. If, on the other hand, $P_{k_2}\psi_2$ is of first type, then one estimates

$$||P_k[P_{k_1}\psi_1 \nabla^{-1} P_{k_2}\psi_2]||_{\dot{X}_k^{0,\frac{1}{2},\infty}} \leq C||P_{k_1}\psi_1||_{A[k_1]+\dot{X}_{k_1}^{0,\frac{1}{2},1}}||P_{k_2}\psi_2||_{A[k_2]} \leq C\tilde{c}_{k_2}$$

with a similar estimate (using (3.5)) controlling the 2nd more complicated norm in the lemma. Summing over $k_2 < -\frac{\epsilon i}{1000C} - 5$ results in the desired exponential gain in $i$. Now assume $k_1 < -\frac{\epsilon i}{1000C} - 5$. In that case, the claim follows from the estimate

$$||P_k[P_{k_1}\psi_1 \nabla^{-1} P_{k_2}\psi_2]||_{\dot{X}_k^{0,\frac{1}{2},1}} \leq C2^{\delta_1(k_1-k)}|k_1-k_2|||P_{k_1}\psi_1||_{S[k_1]}||P_{k_2}\psi_2||_{A[k_2]+\dot{X}_{k_2}^{0,\frac{1}{2},1}},$$

by a calculation similar to the immediately preceding. Summing over $k_1 < -\frac{\epsilon i}{1000C} - 5$ results in the desired exponential gain. Having thus shown that we may assume $|k_1| < \frac{\epsilon i}{1000C} + 5$, we apply the same procedure to the expression $P_{k_1}[\psi_1\nabla^{-1}\psi_2 \ldots \nabla^{-1}\psi_{k-1}]$. Of course the frequency bounds will grow the further one continues this process, but this is counteracted by the rapidly decreasing coefficients coming from the Taylor expansion[8]. Arguing inductively, we see that it suffices to show that provided two functions $\psi_1 \in \mathcal{S}(\mathbf{R}^{2+1})$, $\psi_2 \in \mathcal{S}(\mathbf{R}^{2+1})$ satisfy the assertions in the lemma, then we get the same assertions for the expression

$$P_k\chi_i(t)[P_{k_1}\psi_1\nabla^{-1}P_{k_2}\psi_2]$$

[8]More precisely, one needs to go to expressions of length up to $\frac{\epsilon i}{10000C}$.

where $|k|, |k_1|, |k_2| < \delta i$, $\delta << \epsilon$. First, we estimate

$$\begin{aligned}
&||P_k[P_{k_1}Q_{>k_1-(1-\epsilon)i}[\chi_i(t)\psi_1]\nabla^{-1}P_{k_2}\psi_2]||_{\dot{X}_k^{0,\frac{1}{2},1}} \\
&\leq C(|k|+|k_1|+|k_2|)||P_{k_1}Q_{>k_1-(1-\epsilon)i}[\chi_i(t)\psi_1]||_{\dot{X}_{k_1}^{0,\frac{1}{2},1}}||\nabla^{-1}P_{k_2}\psi_2||_{\dot{X}_{k_2}^{0,\frac{1}{2},1}+A[k_2]} \\
&\leq C|i|2^{-\mu(\epsilon)i}.
\end{aligned}$$

We have used the bound on $P_{k_2}\psi_2$ from theorem 2.3, as well as the easily verified fact that

$$||P_{k_1}Q_{>k_1-(1-\epsilon)i}[\chi_i(t)\psi_1]||_{\dot{X}_{k_1}^{0,\frac{1}{2},1}} \leq C2^{-\mu(\epsilon)i}$$

One proceeds similarly when $P_{k_2}\psi_2$ is replaced by $P_{k_2}Q_{>k_2-(1-\epsilon)i}\psi_2$. Thus it suffices to consider the expression

$$P_k\chi_i(t)[P_{k_1}Q_{<k_1-(1-\epsilon)i}\phi\nabla^{-1}P_{k_2}Q_{<k_2-(1-\epsilon)i}\psi]$$

We may always reduce the modulation of the output to size $> 2^{k-\frac{5\epsilon i}{C}}$, as follows easily from 3.4(a). Now we carry out the following decomposition

$$\begin{aligned}
&P_k\chi_i(t)[P_{k_1}Q^{\pm}_{<k_1-(1-\epsilon)i}\phi\nabla^{-1}P_{k_2}Q_{<k_2-(1-\epsilon)i}\psi] \\
&= P_k\chi_i(t)[\sum_{\kappa\in K_{\frac{\epsilon-1}{2}i}}\chi^c_{i,\mp\kappa}P_{k_1,\kappa}Q^{\pm}_{<k_1-(1-\epsilon)i}\phi\nabla^{-1}P_{k_2}Q_{<k_2-(1-\epsilon)i}\psi] \\
&+ P_k\chi_i(t)[\sum_{\kappa\in K_{\frac{\epsilon-1}{2}i}}\chi_{i,\mp\kappa}P_{k_1,\kappa}Q^{\pm}_{<k_1-(1-\epsilon)i}\phi\nabla^{-1}P_{k_2}Q_{<k_2-(1-\epsilon)i}\psi] \\
&+ P_k\chi_i(t)[\phi_i(t,x)P_{k_1}Q^{\pm}_{<k_1-(1-\epsilon)i}\phi\nabla^{-1}P_{k_2}Q_{<k_2-(1-\epsilon)i}\psi]
\end{aligned} \tag{3.6}$$

In the immediately preceding we let $\chi_{i,\pm\kappa}(t,x)$ localize to two slabs aligned or anti-aligned with $\kappa$ of length $\sim 2^i$ and thickness $\sim 2^{\epsilon i}$ (the complement of $\chi^c_{i,\pm\kappa}$ within the $\frac{2^{\epsilon i}}{C}$-neighborhood of the physical light cone), and we let $\phi_i(t,x)$ smoothly localize to the intersection of the complement of the $\frac{2^{\epsilon i}}{C}$ neighborhood of the light cone and a box of dimensions $\sim 2^i\times 2^i\times 2^i$(finite propagation speed and properties of the Fourier multipliers). We estimate

$$\begin{aligned}
&||P_k\chi_i(t)[\phi_i(t,x)P_{k_1}Q_{<k_1-(1-\epsilon)i}\phi\nabla^{-1}P_{k_2}Q_{<k_2-(1-\epsilon)i}\psi]||_{\dot{X}_k^{0,\frac{1}{2},1}} \\
&\leq C2^{\frac{\max\{k,k_1,k_2\}}{2}}||\phi_i(t,x)\chi_i(t)\nabla^{-1}P_{k_2}Q_{<k_2}\psi||_{L^2_tL^\infty_x}||P_{k_1}Q_{<k_1-(1-\epsilon)i}\phi||_{L^\infty_tL^2_x} \\
&\quad + ||P_k\chi_i(t)[\phi_i(t,x)P_{k_1}Q_{<k_1-(1-\epsilon)i}\phi\nabla^{-1}P_{k_2}Q_{[k_2+(\epsilon-1)i,k_2]}\psi]||_{\dot{X}_k^{0,\frac{1}{2},1}}
\end{aligned}$$

Now the first summand on the right hand side is immediately controlled[9] from Proposition 1.1, while we estimate the 2nd as follows(modulo error terms of order

[9]One may have to shrink the size of $|k|$, $|k_{1,2}|$ if necessary.

of magnitude $2^{-Ni}$):

$$||P_k\chi_i(t)[\phi_i(t,x)P_{k_1}Q_{<k_1-(1-\epsilon)i}\phi\nabla^{-1}P_{k_2}Q_{[k_2+(\epsilon-1)i,k_2]}\psi]||_{\dot{X}_k^{0,\frac{1}{2},1}}$$
$$\leq C2^{\frac{\max\{k,k_1,k_2\}}{2}}||P_k\chi_i(t)[\phi_i(t,x)P_{k_1}Q_{<k_1-(1-\epsilon)i}\phi\nabla^{-1}P_{k_2}Q_{[k_2+(\epsilon-1)i,k_2]}\psi]||_{L_t^2L_x^2}$$
$$\leq C2^{\frac{\max\{k,k_1,k_2\}}{2}}||\phi_i(t,x)||_{L_t^\infty L_x^M}||P_{k_1}Q_{<k_1-(1-\epsilon)i}\phi||_{L_t^M L_x^{2+}}$$
$$||\nabla^{-1}P_{k_2}Q_{[k_2+(\epsilon-1)i,k_2]}\chi_i(t)\psi||_{L_t^{2+}L_x^\infty}$$
$$\leq C2^{\frac{\max\{k,k_1,k_2\}}{2}}C(M)2^{(\frac{1}{M}-\mu)i}$$

This is acceptable if we choose $M$ large enough. Now we proceed to the other two terms in (3.6). Consider the first. We have for $\min\{k_1,k_2,k\}+O(1)>j>k-\frac{5\epsilon i}{C}$

$$||P_kQ_j\chi_i(t)[\sum_{\kappa\in K_{\frac{\epsilon-1}{2}i}}\chi^c_{i,\mp\kappa}P_{k_1,\kappa}Q^{\pm}_{<k_1-(1-\epsilon)i}\phi\nabla^{-1}P_{k_2}Q^{\pm}_{<k_2-(1-\epsilon)i}\psi]||_{\dot{X}_k^{0,\frac{1}{2},1}}$$
$$\leq C2^{\frac{\min\{k,k_1,k_2\}-j}{4}}(\sum_{\tilde{\kappa}\in K_{\frac{j-\min\{k,k_1,k_2\}}{2}}}||P_{k_1,\tilde{\kappa}}\sum_{\kappa\in K_{\frac{\epsilon-1}{2}i}}\chi^c_{i,\mp\kappa}P_{k_1,\kappa}Q^{\pm}_{<k_1-(1-\epsilon)i}\phi||^2_{PW[\pm\tilde{\kappa}]})^{\frac{1}{2}}$$
$$(\sum_{\tilde{\kappa}\in K_{\frac{j-\min\{k,k_1,k_2\}}{2}}}||\nabla^{-1}P_{k_2,\tilde{\kappa}}Q^{\pm}_{<k_2-(1-\epsilon)i}\psi||^2_{NFA^*[\pm\tilde{\kappa}]})^{\frac{1}{2}},$$

which is bounded by $2^{-\mu i}$. One can easily sum over $j\in[k-ai,k+bi]$, obtaining the desired estimate. If $j>\min\{k_1,k_2,k\}+100$, say, then necessarily $k_1=k_2+O(1)=j+O(1)$, since the inputs $\chi^c_{i,\mp\kappa}P_{k_1,\kappa}Q^{\pm}_{<k_1-(1-\epsilon)i}\psi_2$ etc have modulation $<C2^{-\epsilon i}<<\max\{2^k,2^{k_{1,2}}\}$. Then one can concurrently microlocalize the inputs to caps $\kappa_{1,2}\in K_{-100}$ of distance $\sim 1$, and argue just as before. The 2nd term in (3.6) has to be decomposed further as follows:

$$P_k\chi_i(t)[\sum_{\kappa\in K_{\frac{\epsilon-1}{2}i}}\chi_{i,\mp\kappa}P_{k_1,\kappa}Q^{\pm}_{<k_1-(1-\epsilon)i}\phi\nabla^{-1}P_{k_2}Q_{<k_2-(1-\epsilon)i}\psi]$$
$$=P_k\chi_i(t)[\sum_{\kappa\in K_{\frac{\epsilon-1}{2}i}}\chi_{i,\mp\kappa}P_{k_1,\kappa}Q^{\pm}_{<k_1-(1-\epsilon)i}\phi\nabla^{-1}\sum_{\kappa\in K_{\frac{\epsilon-1}{2}i}}\chi^c_{i,\mp\kappa}P_{k_2,\kappa}Q^{\pm}_{<k_2-(1-\epsilon)i}\psi]]$$
$$+P_k\chi_i(t)[\sum_{\kappa\in K_{\frac{\epsilon-1}{2}i}}\chi_{i,\mp\kappa}P_{k_1,\kappa}Q^{\pm}_{<k_1-(1-\epsilon)i}\phi\nabla^{-1}\sum_{\kappa\in K_{\frac{\epsilon-1}{2}i}}\chi_{i,\mp\kappa}P_{k_2,\kappa}Q^{\pm}_{<k_2-(1-\epsilon)i}\psi]]$$

For the first of the immediately preceding terms, we can estimate[10] provided $j<\min\{k,k_1,k_2\}+O(1)$

$$2^{\frac{j}{2}}||P_kQ_j\chi_i(t)[\sum_{\kappa\in K_{\frac{\epsilon-1}{2}i}}\chi_{i,\mp\kappa}P_{k_1,\kappa}Q^{\pm}_{<k_1-(1-\epsilon)i}\phi$$
$$\nabla^{-1}\sum_{\kappa\in K_{\frac{\epsilon-1}{2}i}}\chi^c_{i,\mp\kappa}P_{k_2,\kappa}Q^{\pm}_{<k_2-(1-\epsilon)i}\psi]]||_{L_t^2L_x^2}$$
$$\leq C(\sum_{\kappa\in K_{\frac{\epsilon-1}{2}i}}||P_kQ_j[P_{k_1,\kappa}Q^{\pm}_{<k_1-(1-\epsilon)i}\phi\nabla^{-1}\sum_{\kappa\in K_{\frac{\epsilon-1}{2}i}}\chi^c_{i,\mp\kappa}P_{k_2,\kappa}Q^{\pm}_{<k_2-(1-\epsilon)i}\psi]||^2_{L_t^2L_x^2})^{\frac{1}{2}}$$

---

[10]Exploit the fact that $\chi_{i,\kappa}$ only smears out the Fourier support by about $2^{-\epsilon i}$, up to negligible error terms.

Using elementary geometry and the definition of $PW[\kappa]$, $NFA^*[\kappa]$ etc., this in turn is estimated by

$$\leq C2^{-\frac{j-\min\{k,k_1,k_2\}}{4}}\Big(\sum_{\kappa\in K_{\frac{\epsilon-1}{2}i}}||P_{k_1,\kappa}Q^{\pm}_{<k_1-(1-\epsilon)i}\phi||^2_{NFA^*[\pm\kappa]}\Big)^{\frac{1}{2}}$$
$$\Big(\sum_{\tilde\kappa\in K_{\frac{j-\min\{k,k_1,k_2\}}{2}}}||P_{k,\tilde\kappa}\sum_{\kappa\in K_{\frac{\epsilon-1}{2}i}}\chi^c_{i,\mp\kappa}P_{k_2,\kappa}Q^{\pm}_{<k_2-(1-\epsilon)i}\psi||^2_{PW[\pm\tilde\kappa]}\Big)^{\frac{1}{2}}$$

This can again be estimated by $\leq C2^{-\mu i}$, as desired. The case $j > \min\{k,k_1,k_2\} + O(1)$ is treated as before. Finally, we have

$$||P_kQ_j\chi_i(t)[\sum_{\kappa\in K_{\frac{\epsilon-1}{2}i}}\chi_{i,\mp\kappa}P_{k_1,\kappa}Q^{\pm}_{<k_1-(1-\epsilon)i}\phi$$
$$\nabla^{-1}\sum_{\kappa\in K_{\frac{\epsilon-1}{2}i}}\chi_{i,\mp\kappa}P_{k_2,\kappa}Q^{\pm}_{<k_2-(1-\epsilon)i}\psi]]||_{L^2_tL^2_x}$$
$$\leq C2^{-Ni}$$

on account of the support properties of the (Fourier)multipliers. We deduce from this that the expression $P_k\chi_i(t)[P_{k_1}\phi\nabla^{-1}P_{k_2}\psi]$ also satisfies the 2nd property of the lemma: simply apply (2.3) in addition to (3.5). □

We can now conclude case (I.b.2). The preceding proof in addition to 3.4(g) imply that

$$||P_{k_0}Q_{<k_0}[R_\alpha P_{k_1}Q_{<k_0}\delta\psi_1P_k\triangle^{-1}\sum_{j=1,2}\partial_j\Box I[\nabla^{-1}P_{k_2}\chi_i(t)\psi_2R_jP_{k_3}\psi_3]]||_{N[k_0]}$$
$$\leq C||P_{k_1}\delta\psi_1||_{S[k_1]}||P_kI[\nabla^{-1}P_{k_2}\chi_i(t)\psi_2R_jP_{k_3}\psi_3]||_{\dot X_k^{0,\frac{1}{2},1}} \leq C2^{-\mu i}\tilde c_{k_1}$$

One can sum over the admissible frequency ranges here (picking up factors $O(i)$), which yields the desired result.
(I.b.3): *The third term in the expansion:*

$$P_{k_0}Q_{<k_0}[R_\alpha P_{k_1}Q_{<k_0}\Box\delta\psi_1P_k\triangle^{-1}\sum_{j=1,2}\partial_jI[\nabla^{-1}P_{k_2}\chi_i(t)\psi_2R_jP_{k_3}\psi_3]]$$

This is much simpler to estimate, on account of the strong Strichartz type estimates available for $\psi_\nu$: we have

$$||P_{k_0}Q_{<k_0}[R_\alpha P_{k_1}Q_{<k_0}\Box\delta\psi_1P_k\triangle^{-1}\sum_{j=1,2}\partial_jI[\nabla^{-1}P_{k_2}\chi_i(t)\psi_2R_jP_{k_3}\psi_3]]||_{L^1_t\dot H^{-1}}$$
$$\leq C||P_{k_1}Q_{<k_0}\Box\delta\psi_1||_{L^2_tL^M_x}||\nabla^{-1}P_{k_2}\chi_i(t)\psi_2||_{L^4_tL^{4+}_x}||P_{k_3}\psi_3||_{L^4_tL^{4+}_x} \leq C2^{-\mu i}\tilde c_{k_1}$$

(I.b.4) *The fourth term of the expansion:*

$$P_{k_0}Q_{<k_0}[P_{k_3}Q_{<k_3}\nabla^{-1}\psi_3IP_{k_2+O(1)}\Box[\nabla^{-1}P_{k_2}Q_{<k_2}\chi_i(t)\psi_2R_\alpha P_{k_1}Q_{<k_0}\delta\psi_1]]$$

This is straightforward by means of 3.4(a), 3.4(g): one estimates[11]

$$||P_{k_0}Q_{<k_0}[P_{k_3}Q_{<k_3}\nabla^{-1}\psi_3 P_{k_2+O(1)}I\Box[\nabla^{-1}P_{k_2}Q_{<k_2}\chi_i(t)\psi_2 R_\alpha P_{k_1}Q_{<k_0}\delta\psi_1]]||_{N[k_0]}$$
$$\leq C|k_0-k_3|||P_{k_3}\nabla^{-1}\psi_3||_{A[k_3]+\dot{X}_{k_3}^{0,\frac{1}{2},1}}$$
$$||[\nabla^{-1}P_{k_2}Q_{<k_2}\chi_i(t)\psi_2 R_\alpha P_{k_1}Q_{<k_0}\delta\psi_1]||_{\dot{X}_{k_2}^{1,\frac{1}{2},1}}$$
$$\leq C|k_0-k_3|||P_{k_3}\nabla^{-1}\psi_3||_{\dot{X}_{k_3}^{0,\frac{1}{2},1}+A[k_3]}$$
$$2^{k_1}||P_{k_1}\delta\psi_1||_{S[k_1]}||\nabla^{-1}P_{k_2}Q_{<k_2}\chi_i(t)\psi_2||_{\dot{X}_{k_2}^{0,\frac{1}{2},1}+A[k_2]}$$
$$\leq C2^{-\mu i}\tilde{c}_{k_1}$$

(I.b.5) *The fifth term of the expansion.*

$$P_{k_0}Q_{<k_0}[P_{k_3}Q_{<k_3}\nabla^{-1}\psi_3 P_{k_3}I[\nabla^{-1}\Box P_{k_2}Q_{<k_2+O(1)}\chi_i(t)\psi_2 R_\alpha P_{k_1}Q_{<k_0}\delta\psi_1]]$$

This can be estimated by means of lemma 3.3: observe that we may throw an operator $Q_{[k_2+(\epsilon-1)i,k_2]}$ in front of $P_{k_2}Q_{<k_2+O(1)}\chi_i(t)\psi_2$, since in the opposite case we have

$$||P_{k_0}Q_{<k_0}[P_{k_3}Q_{<k_3}\nabla^{-1}\psi_3 P_{k_3}I[\nabla^{-1}\Box P_{k_2}Q_{<k_2+(\epsilon-1)i}\chi_i(t)\psi_2 R_\alpha P_{k_1}Q_{<k_0}\delta\psi_1]]||_{L_t^1\dot{H}^{-1}}$$
$$\leq C2^{-k_0}||P_{k_3}Q_{<k_3}\nabla^{-1}\psi_3||_{L_t^{2+}L_x^\infty}||\nabla^{-1}\Box P_{k_2}Q_{<k_2+(\epsilon-1)i}\chi_i(t)\psi_2||_{L_t^2L_x^2}||P_{k_1}\delta\psi_1||_{L_t^M L_x^\infty}$$
$$\leq C2^{-k_0}2^{\frac{\epsilon-1}{2}i}2^{(1-\frac{1}{M})k_1}\tilde{c}_{k_1}\leq C2^{-\mu i}\tilde{c}_{k_1}$$

for very large $M$, on account of the assumptions on $k_1, k_0$. Using lemma 3.3 we have

$$||P_{k_0}Q_{<k_0}[R_\alpha P_{k_3}Q_{<k_3}\nabla^{-1}\psi_3 P_{k_3}I[\nabla^{-1}\Box P_{k_2}Q_{[k_2+(\epsilon-1)i,k_2]}\chi_i(t)\psi_2 R_j P_{k_1}\delta\psi_1]]||_{L_t^1\dot{H}^{-1}}$$
$$\leq C2^{-k_0}||P_{k_3}\nabla^{-1}\psi_3||_{L_t^{2+}L_x^\infty}||P_{k_2}Q_{[k_2+(\epsilon-1)i,k_2]}\chi_i(t)\psi_2||_{\dot{X}_{k_2}^{0,\frac{1}{2},1}}||P_{k_1}\delta\psi_1||_{L_t^M L_x^\infty}$$
$$\leq C2^{-\mu i}2^{-\frac{k_1}{M}}\tilde{c}_{k_1}$$

Choosing $M$ large enough results in the desired gain in $i$.
(I.b.6): *The sixth term of the expansion.* This is similar to the third and hence omitted. The terms in the expansion (3.4) are simple variations of the same kind of reasoning and hence omitted. This concludes case (I.b).
(I.c): *None of (I.a), (I.b) hold but $i\leq C|k_1|$.* This implies $|k_{2,3}|, |k| << i$, and we may treat these as $O(1)$. We also have $i\leq Ck_1$, and $k_1 = k_0 + O(1)$. Now we proceed in close analogy to the immediately preceding case. We may pull the multiplier $\chi_i(t)$ past the operator $P_k Q I\nabla^{-1}$ right in front of $P_{k_2}\psi_2$. Also, we may reduce the output and first input $P_{k_1}\delta\psi_1$ to modulation $< O(1)$: for example, one estimates

$$||P_{k_0}Q_{\geq 0}\nabla_{x,t}[P_{k_1}Q_{<k_1}R_\alpha\delta\psi_1\nabla^{-1}P_k I[P_{k_2}\chi_i(t)\psi_2 P_{k_3}\psi_3]]||_{\dot{X}_{k_0}^{-1,-\frac{1}{2},1}}$$
$$\leq C||P_{k_1}\delta\psi_1||_{L_t^\infty L_x^2}||\nabla^{-1}P_k I[P_{k_2}\chi_i(t)\psi_2 P_{k_3}\psi_3]||_{L_t^2L_x^\infty}$$
$$\leq C2^{-\mu i}\tilde{c}_{k_1}$$

[11]Recall the definition of $S[k]$ via $||.||_{A[k]}$, $||.||_{B[k]}$. Also recall that we may replace $P_{k_2}\psi_2$ by $P_{k_2}Q_{<k_2}\psi_2$ etc.

on account of the proof of lemma 3.3, and the fact that by theorem 2.1 we have

$$||\nabla^{-1}P_kI[P_{k_2}\chi_i(t)\psi_2P_{k_3}\psi_3]||_{L_t^2L_x^\infty}\leq C||\nabla^{-1}P_kI[P_{k_2}\chi_i(t)\psi_2P_{k_3}\psi_3]||_{\dot{X}_k^{\epsilon,\frac{1}{2+},1}}$$

The estimate for when $P_{k_1}R_\alpha\delta\psi_1$ is replaced by $P_{k_1}Q_{\geq k_1}R_\alpha\psi_1$ is of course similar, since one may place this into $L_t^2L_x^2$ and gains $2^{-\frac{k_1}{2}}$. Then we resort to the null-form identities (3.3), (3.4). We wind up having to estimate the same types of expressions as in (I.b.1-6), with our changed assumptions on the frequencies. We shall do this in a brisk manner here:

(I.c.1): we need to exert some care not to lose in $k_0$: use the proof of lemma 3.3 as above to obtain

$$||P_{k_0}Q_{<O(1)}\Box[R_\alpha P_{k_1}Q_{<O(1)}\delta\psi_1P_k\triangle^{-1}\sum_{j=1,2}\partial_jI[\nabla^{-1}P_{k_2}\chi_i(t)\psi_2R_jP_{k_3}\psi_3]]||_{\dot{X}_{k_0}^{-1,-\frac{1}{2},1}}$$

$$\leq C||P_{k_1}Q_{<O(1)}\delta\psi_1||_{L_t^\infty L_x^2}||P_k\triangle^{-1}\sum_{j=1,2}\partial_jI[\nabla^{-1}P_{k_2}\chi_i(t)\psi_2R_jP_{k_3}\psi_3]||_{L_t^2L_x^\infty}$$

$$\leq C||P_{k_1}Q_{<O(1)}\delta\psi_1||_{L_t^\infty L_x^2}||P_k\triangle^{-1}\sum_{j=1,2}\partial_jI[\nabla^{-1}P_{k_2}\chi_i(t)\psi_2R_jP_{k_3}\psi_3]||_{\dot{X}_k^{\epsilon,\frac{1}{2+},1}}$$

$$\leq C2^{-\mu i}\tilde{c}_{k_0}$$

(I.c.2): This is estimated exactly like (I.b.2).

(I.c.3): The argument is here is slightly more complicated than in (I.b.3); we argue as in (I.c.2):

$$||P_{k_0}Q_{<O(1)}[R_\alpha P_{k_1}Q_{<O(1)}\Box\delta\psi_1P_k\triangle^{-1}\sum_{j=1,2}\partial_jI[\nabla^{-1}P_{k_2}\chi_i(t)\psi_2R_jP_{k_3}\psi_3]]||_{L_t^1\dot{H}^{-1}}$$

$$\leq C2^{-k_0}||P_{k_1}Q_{<O(1)}\Box\delta\psi_1||_{L_t^2L_x^2}||P_k\triangle^{-1}\sum_{j=1,2}\partial_jI[\nabla^{-1}P_{k_2}\chi_i(t)\psi_2R_jP_{k_3}\psi_3]||_{L_t^2L_x^\infty}$$

$$\leq C\tilde{c}_{k_0}||P_k\triangle^{-1}\sum_{j=1,2}\partial_jI[\nabla^{-1}P_{k_2}\chi_i(t)\psi_2R_jP_{k_3}\psi_3]||_{\dot{X}_k^{\epsilon,\frac{1}{2+},1}}\leq C2^{-\mu i}\tilde{c}_{k_0}$$

(I.c.4): The argument here is more complicated than in (I.b.4) on account of the fact that we don't get the desired exponential gain from simple application of 3.4(a). Instead, we shall have to resort to lemma 3.3. We split the expression into several manageable pieces: first, using 3.4(g), we can estimate[12]

$$\begin{aligned}&||P_{k_0}Q_{<O(1)}[P_{k_3}Q_{<k_3}\nabla^{-1}\psi_3\\&\qquad I\Box[\nabla^{-1}P_{k_2}Q_{[k_2+(\epsilon-1)i,k_2+O(1)]}(\chi_i(t)\psi_2)R_jP_{k_1}R_\alpha\delta\psi_1]]||_{N[k_0]}\\&\leq C||P_{k_3}\psi_3\nabla^{-1}\psi_3||_{\dot{X}_{k_3}^{0,\frac{1}{2},1}+A[k_3]}\\&\qquad||\nabla^{-1}P_{k_2}Q_{[k_2+(\epsilon-1)i,k_2+O(1)]}(\chi_i(t)\psi_2)||_{\dot{X}_{k_2}^{0,\frac{1}{2},1}}||P_{k_1}R_\alpha\delta\psi_1||_{S[k_3]}\leq C2^{-\mu i}\tilde{c}_{k_0}\end{aligned}$$

Next we reduce the large-frequency input $P_{k_1}\delta\psi_1$ to modulation $< 2^{-i\delta}$. For this, throw an operator $Q_{\geq-i\delta}$ in front of $P_{k_1}\delta\psi_1$, let the operator $\Box$ fall inside the inner square bracket and use the proof of lemma 3.3, placing the products $P_{k_3}Q_{<k_3}\nabla^{-1}\psi_3\nabla^{-1}P_{k_2}Q_{<k_2+(\epsilon-1)i}(\chi_i(t)\psi_2)$ etc. into $L_t^2L_x^\infty$ and $P_{k_1}Q_{\geq-i\delta}\delta\psi_1$ into $L_t^2L_x^2$ and the output into $L_t^1\dot{H}^{-1}$. Next, we claim that we may apply an operator

[12]It is easy to see that we may restrict $P_{k_1}R_\alpha\delta\psi_1$ to modulation $< O(1)$, so that we don't lose from $R_\alpha$. We shall omit this to streamline the formulae.

$Q_{\geq -i\delta+C}$ in front of the inner square bracket. Indeed, if we apply an operator $Q_{<-i\delta+C}$, we estimate using 3.4(a) that

$$\begin{aligned}
&||P_{k_0}Q_{<O(1)}[P_{k_3}Q_{<k_3}\nabla^{-1}\psi_3\\
&\qquad I\Box Q_{<-i\delta+C}[\nabla^{-1}P_{k_2}Q_{<k_2+(\epsilon-1)i}(\chi_i(t)\psi_2)R_\alpha P_{k_1}Q_{<-i\delta}\delta\psi_1]]||_{N[k_0]}\\
&\leq C||P_{k_3}\psi_3||_{\dot{X}_{k_3}^{0,\frac{1}{2},1}+A[k]}\\
&\qquad ||\Box Q_{<-i\delta+C}[\nabla^{-1}P_{k_2}Q_{<k_2+(\epsilon-1)i}(\chi_i(t)\psi_2)R_j P_{k_1}Q_{<-i\delta}\delta\psi_1]]||_{\dot{X}_{k_1}^{0,-\frac{1}{2},1}}\\
&\leq C2^{\frac{-i\delta-k_2}{4+}}\tilde{c}_{k_1},
\end{aligned}$$

which is acceptable provided $|k_2|$ is small enough in relation to $i\delta$, which we may always arrange. Thus we now reduce to estimating

$$\begin{aligned}
&P_{k_0}Q_{<O(1)}[P_{k_3}Q_{<k_3}\nabla^{-1}\psi_3\\
&\qquad I\Box Q_{\geq -i\delta+C}[\nabla^{-1}P_{k_2}Q_{<k_2+(\epsilon-1)i}(\chi_i(t)\psi_2)R_\alpha P_{k_1}Q_{<-i\delta}\delta\psi_1]]
\end{aligned}$$

We note that on account of lemma 3.3, we may estimate[13]

$$\begin{aligned}
&||P_{k_0}Q_{<O(1)}[P_{k_3}Q_{<k_3}\nabla^{-1}\psi_3\\
&I\Box Q_{\geq -i\delta+C}[\nabla^{-1}\sum_{\kappa\in K_{\frac{\epsilon-1}{2}i}}\chi^c_{i,\mp\kappa}P_{k_2,\kappa}Q^{\pm}_{<k_2+(\epsilon-1)i}(\chi_i(t)\psi_2)R_\alpha P_{k_1}Q_{<-i\delta}\delta\psi_1]]||_{N[k_0]}\\
&\leq C||P_{k_3}\psi_3||_{S[k_3]}||I\Box Q_{\geq -i\delta+C}[\nabla^{-1}\sum_{\kappa\in K_{\frac{\epsilon-1}{2}i}}\chi^c_{i,\mp\kappa}P_{k_2,\kappa}Q^{\pm}_{<k_2+(\epsilon-1)i}(\chi_i(t)\psi_2)\\
&\qquad\qquad R_\alpha P_{k_1}Q_{<-i\delta}\delta\psi_1]]||_{\dot{X}_{k_1}^{0,-\frac{1}{2},1}},
\end{aligned}$$

and one has (see the arguments in [**23**])

$$\begin{aligned}
&||P_{k_1+O(1)}Q_{\geq -i\delta+C}[\nabla^{-1}\sum_{\kappa\in K_{\frac{\epsilon-1}{2}i}}\chi^c_{i,\mp\kappa}P_{k_2,\kappa}Q^{\pm}_{<k_2+(\epsilon-1)i}(\chi_i(t)\psi_2)\\
&\qquad\qquad R_j P_{k_1}Q_{<-i\delta}\delta\psi_1]]||_{\dot{X}_{k_1}^{0,\frac{1}{2},1}},\\
&\leq C(\sum_{\tilde{\kappa}\in K_{\frac{-i\delta+C-k_2}{2}}}||\nabla^{-1}P_{k_2,\tilde{\kappa}}\sum_{\kappa\in K_{\frac{\epsilon-1}{2}i}}\chi^c_{i,\mp\kappa}P_{k_2,\kappa}Q^{\pm}_{<k_2+(\epsilon-1)i}(\chi_i(t)\psi_2)||^2_{PW[\pm\tilde{\kappa}]})^{\frac{1}{2}}\\
&\qquad\qquad 2^{\frac{-\delta i-k_2}{4}}(\sum_{\tilde{\kappa}\in K_{\frac{-i\delta+C-k_2}{2}}}||P_{k_1,\tilde{\kappa}}Q^{\pm}_{<-i\delta}\delta\psi_1||^2_{NFA^*[\pm\tilde{\kappa}]})^{\frac{1}{2}}\\
&\leq C2^{-\mu i}\tilde{c}_{k_1}
\end{aligned}$$

---

[13] We need to assume that $\delta << \epsilon$, which we may always arrange.

Also, arguing as in the proof of lemma 3.3 and using the same notation, we can estimate

$$||P_{k_0}Q_{<O(1)}[P_{k_3}Q_{<k_3}\nabla^{-1}\psi_3$$
$$I\Box Q_{\geq -i\delta+C}[\phi_i(t,x)\nabla^{-1}P_{k_2}Q_{<k_2+(\epsilon-1)i}(\chi_i(t)\psi_2)R_\alpha P_{k_1}Q_{<-i\delta}\delta\psi_1]]||_{L_t^1\dot{H}^{-1}}$$
$$\leq C||\phi_i(t,x)P_{k_3}Q_{<k_3}\nabla^{-1}\psi_3||_{L_t^2L_x^\infty}||\phi_i(t,x)\nabla^{-1}P_{k_2}Q_{<k_2}\chi_i(t)\psi_2||_{L_t^2L_x^\infty}||P_{k_1}\delta\psi_1||_{L_t^\infty L_x^2}$$
$$+\,||\phi_i(t,x)R_\alpha P_{k_3}Q_{<k_3}\nabla^{-1}\psi_3||_{L_t^2L_x^\infty}||\phi_i(t,x)||_{L_t^ML_x^\infty}$$
$$||\nabla^{-1}P_{k_2}Q_{[k_2+(\epsilon-1)i,k_2]}(\chi_i(t)\psi_2)||_{L_t^{2+}L_x^\infty}||P_{k_1}\delta\psi_1||_{L_t^\infty L_x^2}$$
$$\leq C2^{-\mu i}\tilde{c}_{k_1}$$

This means we only need to estimate the following expression:

$$P_{k_0}Q_{<O(1)}[P_{k_3}Q_{<k_3}\nabla^{-1}\psi_3$$
$$I\Box Q_{\geq -i\delta+C}[\nabla^{-1}\sum_{\kappa\in K_{\frac{\epsilon-1}{2}i}}\chi_{i,\mp\kappa}P_{k_2,\kappa}Q^{\pm}_{<k_2+(\epsilon-1)i}(\chi_i(t)\psi_2)R_\alpha P_{k_1}Q_{<-i\delta}\delta\psi_1]]$$

Using 3.4(g), it suffices to estimate

$$||I\Box Q_{\geq -i\delta+C}[\nabla^{-1}\sum_{\kappa\in K_{\frac{\epsilon-1}{2}i}}\chi_{i,\mp\kappa}P_{k_2,\kappa}Q^{\pm}_{<k_2+(\epsilon-1)i}(\chi_i(t)\psi_2)R_\alpha P_{k_1}Q_{<-i\delta}\delta\psi_1]]||_{\dot{X}_{k_1}^{0,-\frac{1}{2},1}}$$

We have the identity[14]

$$Q_{\geq -i\delta+C}[\nabla^{-1}\sum_{\kappa\in K_{\frac{\epsilon-1}{2}i}}\chi_{i,\mp\kappa}\nabla^{-1}P_{k_2,\kappa}Q^{\pm}_{<k_2+(\epsilon-1)i}(\psi_2)R_\alpha P_{k_1}Q^{\pm}_{<-i\delta}\delta\psi_1]$$
$$=\sum_{\tilde{\kappa}_{1,2}\in K_{\frac{-\delta i-k_2}{2}},\mathrm{dist}(\pm\kappa_1,\pm\kappa_2)\sim 2^{\frac{-i\delta-k_2}{2}}}$$
$$Q_{\geq -i\delta+C}[P_{k_2,\tilde{\kappa_1}}\nabla^{-1}\sum_{\kappa\in K_{\frac{\epsilon-1}{2}i}}\chi_{i,\mp\kappa}\nabla^{-1}P_{k_2,\kappa}Q^{\pm}_{<k_2+(\epsilon-1)i}(\psi_2)R_\alpha P_{k_1,\tilde{\kappa_2}}Q^{\pm}_{<-i\delta}\delta\psi_1]$$

One can move the multiplier $\chi_{i,\kappa}$ to the outside of this expression while generating error terms rapidly decaying in $i$ and hence negligible for the argument. Hence exploiting orthogonality we may now estimate

$$||Q_{\geq -i\delta+C}[\sum_{\kappa\in K_{\frac{\epsilon-1}{2}i}}\chi_{i,\pm\kappa}\nabla^{-1}P_{k_2,\kappa}Q^{\pm}_{<k_2+(\epsilon-1)i}(\psi_2)R_\alpha P_{k_1}Q^{\pm}_{<-i\delta}\delta\psi_1]]||_{L_t^2L_x^2}$$
$$\leq C(\sum_{\kappa\in K_{\frac{\epsilon-1}{2}i}}[\sum_{\tilde{\kappa}_{1,2}\in K_{\frac{-\delta i-k_2}{2}},\mathrm{dist}(\pm\tilde{\kappa}_1,\pm\tilde{\kappa}_2)\sim 2^{\frac{-i\delta-k_2}{2}},\kappa\cap\tilde{\kappa}_1\neq\emptyset}$$
$$||Q_{\geq -i\delta+C}[P_{k_2,\pm\tilde{\kappa_1}}\nabla^{-1}P_{k_2,\kappa}Q^{\pm}_{<k_2+(\epsilon-1)i}\psi_2R_\alpha P_{k_1,\tilde{\kappa_2}}Q^{\pm}_{<-i\delta}\delta\psi_1]]||_{L_t^2L_x^2}]^2)^{\frac{1}{2}}$$

On the other hand, using (2.1), we can easily estimate that

$$||Q_{\geq -i\delta+C}[P_{k_2,\tilde{\kappa}_1}\nabla^{-1}P_{k_2,\kappa}Q^{\pm}_{<k_2+(\epsilon-1)i}\psi_2R_\alpha P_{k_1,\tilde{\kappa_2}}Q^{\pm}_{<-i\delta}\delta\psi_1]]||_{L_t^2L_x^2}$$
$$\leq C2^{-\frac{-i\delta-k_2}{4}}||\nabla^{-1}P_{k_2,\kappa}Q^{\pm}_{<k_2+(\epsilon-1)i}\psi_2||_{PW[\pm\kappa]}||P_{k_1,\tilde{\kappa_2}}Q^{\pm}_{<-i\delta}\delta\psi_1||_{NFA[\pm\tilde{\kappa}_2]}$$

---

[14]Again this requires that $\delta << \epsilon$.

Plugging this back into the preceding inequality and using the definition of $S[k,\kappa]$, we arrive at the following:

$$\begin{aligned}
&||Q_{\geq -i\delta+C}[\sum_{\kappa\in K_{\frac{\epsilon-1}{2}i}} \chi_{i,\mp\kappa}\nabla^{-1}P_{k_2,\kappa}Q^{\pm}_{<k_2+(\epsilon-1)i}(\psi_2)R_\alpha P_{k_1}Q^{\pm}_{<-i\delta}\delta\psi_1]||_{L_t^2L_x^2}\\
&\leq C2^{\frac{\epsilon-1}{4}i}2^{-\frac{-i\delta-k_2}{4}}(\sum_{\kappa\in K_{\frac{\epsilon-1}{2}i}}||\nabla^{-1}P_{k_2,\kappa}Q^{\pm}_{<k_2+(\epsilon-1)i}\psi_2||^2_{S[k_2,\pm\kappa_2]})^{\frac{1}{2}}\\
&\qquad\qquad(\sum_{\tilde\kappa\in K_{\frac{-\delta i-k_2}{2}}}||P_{k_1,\tilde\kappa}Q^{\pm}_{<-i\delta}\delta\psi_1||^2_{S[k_1,\pm\tilde\kappa]})^{\frac{1}{2}}
\end{aligned}$$

It follows easily that this expression is $\leq C2^{-\mu i}\tilde{c}_{k_1}$, which is as desired.
(I.c.5): This is much simpler since a derivative has been moved from a high-frequency term to a low-frequency term. We use finite propagation speed and Hoelder's inequality to conclude

$$\begin{aligned}
&||P_{k_0}Q_{<O(1)}[P_{k_3}Q_{<k_3}\nabla^{-1}\psi_3\nabla^{-1}\Box[P_{k_2}Q_{<k_2}\psi_2]R_\alpha P_{k_1}\delta\psi_1]||_{L_t^1\dot{H}^{-1}}\\
&\leq C2^{-k_0}2^{\delta i}||P_{k_3}Q_{<k_3}\nabla^{-1}\psi_3||_{L_t^{2+}L_x^\infty}||\nabla^{-1}\Box[P_{k_2}Q_{<k_2}\psi_2]||_{L_t^2L_x^\infty}||R_\alpha P_{k_1}\delta\psi_1||_{L_t^ML_x^{2+}}\\
&\leq C2^{\delta i-k_0}\tilde{c}_{k_1}
\end{aligned}$$

One can choose $\delta=\delta(M)$ arbitrarily small so one gets the desired exponential gain in $i$.
(I.c.6): This is the expression

$$\nabla^{-1}P_{k_2}Q_{<k_2+O(1)}\psi_2\nabla^{-1}P_{k_3}\psi_3\Box P_{k_1}Q_{<O(1)}\delta\psi_1$$

One estimates this using 3.4(g) as well as lemma 3.3:

$$\begin{aligned}
&||P_{k_0}Q_{<O(1)}[\nabla^{-1}P_{k_2}Q_{<k_2+O(1)}\psi_2\nabla^{-1}P_{k_3}\psi_3\Box P_{k_1}Q_{<O(1)}\delta\psi_1]||_{N[k_0]}\\
&\leq C2^{k-k_1}\sum_{k<\max\{k_2,k_3\}+O(1)}||P_k(\nabla^{-1}P_{k_2}Q_{<k_2+O(1)}\psi_2\nabla^{-1}P_{k_3}\psi_3)||_{\dot{X}_k^{0,\frac{1}{2},1}}\\
&\qquad\qquad||\Box P_{k_1}Q_{<O(1)}\delta\psi_1||_{\dot{X}_{k_1}^{0,-\frac{1}{2},1}}\\
&\leq C2^{-\mu i}\tilde{c}_{k_1}
\end{aligned}$$

The terms on the right-hand side of (3.4) are treated analogously. This concludes case (I.c).
(I.d): *None of (I.a), (I.b), (I.c) hold.* In that case we may assume $|k_{1,2,3}|=O(1)$, $|k|=O(1)$. One proceeds exactly as in the immediately preceding case (I.c), but has to argue differently for case (I.d.5): In that case, we split the expression into two manageable pieces:

$$\begin{aligned}
&||P_{k_0}[P_{k_3}Q_{<k_3}\nabla^{-1}\psi_3\nabla^{-1}\Box[P_{k_2}Q_{[k_2+(\epsilon-1)i,k_2]}\chi_i(t)\psi_2]R_\alpha P_{k_1}Q_{<O(1)}\delta\psi_1]||_{N[k_0]}\\
&\leq C\sum_{k<\max\{k_3,k_1\}+O(1)}2^{k-k_2}||R_\alpha P_{k_3}Q_{<k_3}\nabla^{-1}\psi_3R_\alpha P_{k_1}Q_{<k_1}\delta\psi_1||_{\dot{X}_k^{0,\frac{1}{2},1}}\\
&\qquad\qquad||\Box[P_{k_2}Q_{[k_2+(\epsilon-1)i,k_2]}\chi_i(t)\psi_2]||_{\dot{X}_{k_2}^{0,-\frac{1}{2},1}}\\
&\leq C2^{-\mu i}\tilde{c}_{k_1}
\end{aligned}$$

We have used here 3.4(g), 3.4(a), as well as lemma 3.3. Next, once $P_{k_2}\psi_2$ is reduced to very low modulation, we can estimate

$$
\begin{aligned}
&||P_{k_0}[P_{k_3}Q_{<k_3}\nabla^{-1}\psi_3\nabla^{-1}\Box[P_{k_2}Q_{<k_2+(\epsilon-1)i}\chi_i(t)\psi_2]R_\alpha P_{k_1}Q_{<O(1)}\delta\psi_1]||_{L_t^1\dot{H}^{-1}}\\
&\leq C||P_{k_3}\nabla^{-1}\psi_3||_{L_t^{2+}L_x^\infty}||\nabla^{-1}\Box[P_{k_2}Q_{<k_2+(\epsilon-1)i}\chi_i(t)\psi_2]||_{L_t^2L_x^2}||P_{k_1}\delta\psi_1||_{L_t^M L_x^{2+}}\\
&\leq C2^{-\mu i}\tilde{c}_{k_1}
\end{aligned}
$$

This concludes the treatment of case (I).

(II): *The term* **(III)**. This is treated analogously to case (I). One invokes the identity

$$
\begin{aligned}
&\partial^\nu[R_\nu h\sum_{j=1,2}\triangle^{-1}\partial_j[R_ifR_jg-R_jfR_ig]]\\
&=\frac{1}{2}\Box\nabla^{-1}h\sum_{j=1,2}\triangle^{-1}\partial_j[R_ifR_jg-R_jfR_ig]\\
&+\frac{1}{2}\Box[\nabla^{-1}h\sum_{j=1,2}\triangle^{-1}\partial_j[R_ifR_jg-R_jfR_ig]]\\
&-\frac{1}{2}[\nabla^{-1}h\sum_{j=1,2}\triangle^{-1}\partial_j\Box[R_ifR_jg-R_jfR_ig]]
\end{aligned}
$$

(III): *The sum of terms* **(II+VI)**. We frequency localize the expression and its inputs as in the preceding. If both $P_{k_1}\psi_1$, $P_{k_3}\psi_3$ are of the first type, we may assume that $|k_j| << i$, $|k| << i$, due to lemma 2.6 as well as the trilinear estimates 3.4(c). Now assume at least one of them is of the 2nd kind. Then the estimate is straightforward on account of the strong Lebesgue type estimates available: for example, assume that $P_{k_1}\psi_1$ is of the 2nd type. We need to estimate the schematically written expression[15]

$$\nabla_{x,t}P_{k_0}[P_{k_1}R_\nu\psi_1\nabla^{-1}P_kI[P_{k_2}\delta\psi_2P_{k_3}\psi_3]]$$

In this case one has the estimate

$$
\begin{aligned}
&||\nabla_{x,t}P_{k_0}[P_{k_1}R_\nu\psi_1\nabla^{-1}P_kI[P_{k_2}\delta\psi_2P_{k_3}\psi_3]]||_{N[k_0]}\\
&\leq C2^{-\delta_1|k_1-k_0|}2^{\delta_2[\min\{k,k_{2,3}\}-\max\{k,k_{2,3}\}]}\tilde{c}_{k_0}
\end{aligned}
$$

This is a simple exercise with the exception of the case when $\nu=0$ and $P_{k_1}R_0\psi_1$ has large modulation, i. e. we replace this expression by $P_{k_1}Q_{>k_1+100}R_0\psi_1$. Assume that $k_1 >> k$, the other situations being similar or simpler. Then we have

$$
\begin{aligned}
&||\nabla_{x,t}P_{k_0}[P_{k_1}Q_{>k_1+100}R_0\psi_1\nabla^{-1}P_kI[P_{k_2}\delta\psi_2P_{k_3}\psi_3]]||_{N[k_0]}\\
&\leq C2^{-\frac{k_1}{2}}||P_{k_0}[P_{k_1}Q_{>k_1+100}R_0\psi_1\nabla^{-1}P_kI[P_{k_2}\delta\psi_2P_{k_3}\psi_3]]||_{L_t^2L_x^2}\\
&\leq C2^{-\frac{k_1}{2}}||P_{k_0}[P_{k_1}Q_{>k_1+100}R_0\psi_1||_{L_t^\infty L_x^2}||\nabla^{-1}P_kI[P_{k_2}\delta\psi_2P_{k_3}\psi_3]]||_{L_t^2L_x^\infty}\\
&\leq C2^{\frac{k-k_1}{2}}\epsilon2^{\frac{\min\{k,k_2,k_3\}-\max\{k,k_2,k_3\}}{2}}\frac{\tilde{c}_{k_2}}{\epsilon}\frac{\tilde{c}_{k_3}}{\epsilon}.
\end{aligned}
$$

Thus in that case, too, we may assume that $\max_{j=0,\ldots,3}\{|k|,|k_j|\} << i$. Next, we claim that we may also reduce the modulations of all inputs to size $< 1$. Indeed,

[15]It is to be kept in mind that the inner square bracket stands for a null-form of type $Q_{\nu j}$.

for example assume that the first input $P_{k_1}\psi_1$ has modulation $> 1$. Then we may estimate (using schematic notation)

$$||P_{k_0}Q_{<k_0}\nabla_{x,t}\chi_i(t)[P_{k_1}Q_{>0}R_\nu\psi_1\nabla^{-1}P_kI[P_{k_2}\delta\psi_2, P_{k_3}\psi_3]]||_{L_t^1\dot{H}^{-1}}$$
$$\leq C||P_{k_1}Q_{[0,\max\{k_2,k_3\}]}\psi_1||_{\dot{X}_{k_1}^{-\frac{1}{2},1,2}}||P_{k_2}\delta\psi_2||_{L_t^M L_x^{2+}}||P_{k_3}\chi_i(t)\psi_3||_{L_t^{2+}L_x^\infty} \leq C2^{-\mu i}\tilde{c}_{k_0}$$

If, on the other hand, we replace $P_{k_0}Q_{<k_0}$ by $P_{k_0}Q_{\geq k_0}$, we have to be careful when $P_{k_1}\psi_1$ is of the 2nd kind. Then we first reduce both $P_{k_2}\delta\psi_2$, $P_{k_3}\psi_3$ to modulation $< 2^{\delta i+O(1)}$ (which is straightforward), and estimate

$$||P_{k_0}Q_{\geq k_0}\nabla_{x,t}\chi_i(t)[P_{k_1}Q_{>0}R_\nu\psi_1\nabla^{-1}P_kI[P_{k_2}Q_{<\delta i}\delta\psi_2, P_{k_3}Q_{<\delta i}\psi_3]]||_{\dot{X}_{k_0}^{-\frac{1}{2},-1,2}}$$
$$\leq C2^{-k}||P_{k_1}Q_{>0}R_\nu\psi_1||_{L_t^\infty L_x^2}||\nabla_{x,t}\nabla^{-1}P_{k_2}Q_{<\delta i}\delta\psi_2||_{L_t^M L_x^{2+}}$$
$$||\chi_i(t)P_{k_3}Q_{<\delta i}\nabla_{x,t}\nabla^{-1}\psi_3||_{L_t^{2+}L_x^\infty}$$

Assuming (as we may) that all the absolute frequencies $|k|$, $|k_i|$ as well as $\delta i$ are much less than $(\frac{1}{2}-\frac{1}{2+})i$ and summing over these frequency ranges, one obtains from this and the definition of frequency envelope the upper bound $\leq C\tilde{c}_{k_0}2^{-\mu i}$, which is as desired. The remaining cases are handled similarly. Now one reverts to the expansions (3.3), (3.4). The only term requiring different treatment than in case (I) is the 2nd term of the expansion. We record this with the appropriate microlocalizations as follows:

$$P_{k_0}\chi_i(t)[P_{k_1}Q_{<0}\psi_1\sum_{j=1,2}\triangle^{-1}\partial_j\Box IP_k[\nabla^{-1}P_{k_2}\delta\psi_2P_{k_3}\psi_3]]$$

As before, we may innocuously move the localizer $\chi_i(t)$ right in front of the inner square bracket $[,]$. Then we decompose this term as a sum of manageable expressions, keeping in mind lemma 3.3, as well as the estimates in 3.4(a)-3.4(g): first, we have

$$||P_{k_0}\chi_i(t)[P_{k_1}Q_{[k_1+(\epsilon-1)i,k_1]}\psi_1\sum_{j=1,2}\triangle^{-1}\partial_j\Box IP_k[\nabla^{-1}P_{k_2}\delta\psi_2P_{k_3}\psi_3]]||_{N[k_0]}$$
$$\leq C||P_{k_1}Q_{[k_1+(\epsilon-1)i,k_1]}\chi_i(t)\psi_1||_{\dot{X}_{k_1}^{0,\frac{1}{2},1}}||I[\nabla^{-1}P_{k_2}\delta\psi_2P_{k_3}\psi_3]||_{\dot{X}_k^{0,\frac{1}{2},1}} \leq C2^{-\mu i}\tilde{c}_{k_0}$$

Similarly, one estimates

$$||P_{k_0}\chi_i(t)[P_{k_1}Q_{<(\epsilon-1)i+k_1}\psi_1$$
$$\sum_{j=1,2}\triangle^{-1}\partial_j\Box IP_k[\nabla^{-1}P_{k_2}\delta\psi_2P_{k_3}Q_{[k_3+(\epsilon-1)i,k_3]}\psi_3]]||_{N[k_0]}$$
$$\leq C2^{-\mu i}\tilde{c}_{k_0}$$

It is further easy to see that one may reduce the output to modulation $< 2^{-i\delta}$ for some very small[16] $\delta > 0$ (in other words, throw an operator $Q_{<-i\delta}$ in front of it), and also reduce the inner square bracket $\Box IP_k[...]$ to modulation $> 2^{-i\delta+C}$. Further, one easily reduces the input $P_{k_2}\delta\psi_2$ to modulation $< 2^{-i\delta}$. One can then

[16]As usual, we choose $\delta << \epsilon$.

rewrite the expression as follows:

(3.7)

$$
\begin{aligned}
&P_{k_0}Q_{<-i\delta}[\chi_i(t)[P_{k_1}Q_{<(\epsilon-1)i+k_1}\psi_1\\
&\qquad\sum_{j=1,2}\triangle^{-1}\partial_j\Box IP_kQ_{>-i\delta+C}[\nabla^{-1}P_{k_2}Q_{<-i\delta}\delta\psi_2P_{k_3}Q_{<k_3+(\epsilon-1)i}\psi_3]]]\\
&=\sum_{\pm,\pm}\ \sum_{\kappa_{1,2}\in K_{\frac{-i\delta-\min\{k_1,k\}}{2}},\mathrm{dist}(\pm\kappa_1,\pm\kappa_2)\sim 2^{\frac{-i\delta-\min\{k_1,k\}}{2}}}\\
&P_{k_0,\kappa_1}Q^{\pm}_{<-i\delta}\chi_i(t)[P_{k_1,\kappa_2}Q^{\pm}_{<(\epsilon-1)i+k_1}\psi_1\\
&\qquad\sum_{j=1,2}\triangle^{-1}\partial_j\Box IP_kQ_{>-i\delta+C}[\nabla^{-1}P_{k_2}Q_{<-i\delta}\delta\psi_2P_{k_3}Q_{<k_3+(\epsilon-1)i}\psi_3]]
\end{aligned}
$$

Now we have achieved the kind of situation in which the 2nd part of lemma 3.3 becomes useful: indeed, we can estimate

$$
\begin{aligned}
&||P_{k_0}Q_{<-i\delta}\chi_i(t)[\sum_{\kappa\in K_{\frac{\epsilon-1}{2}i}}\chi^c_{i,\mp\kappa}P_{k_1,\kappa}Q^{\pm}_{<(\epsilon-1)i+k_1}\psi_1\\
&\qquad\sum_{j=1,2}\triangle^{-1}\partial_j\Box IP_kQ_{>-i\delta+C}[\nabla^{-1}P_{k_2}Q_{<-\delta i}\delta\psi_2P_{k_3}Q_{<k_3+(\epsilon-1)i}\psi_3]]||_{N[k_0]}\\
&\leq C(\sum_{\kappa_1\in K_{\frac{-i\delta-\min\{k_1,k\}}{2}}}||P_{k_1,\kappa_1}\sum_{\kappa\in K_{\frac{\epsilon-1}{2}i}}\chi^c_{i,\mp\kappa}P_{k_1,\kappa}Q^{\pm}_{<(\epsilon-1)i+k_1}\psi_1||^2_{PW[\kappa_1]})^{\frac{1}{2}}\\
&\qquad\qquad||P_{k_2}Q_{<-\delta i}\delta\psi_2||_{S[k_2]}||P_{k_3}\psi_3||_{A[k_3]+\dot{X}^{0,\frac{1}{2},1}_{k_3}}
\end{aligned}
$$

This furnishes the desired estimate $\leq C2^{-\mu i}\tilde{c}_{k_0}$. One similarly treats the contribution when $P_{k_3}\psi_3$ is replaced by $\sum_{\pm}\sum_{\kappa\in K_{\frac{\epsilon-1}{2}i}}\chi^c_{i,\mp\kappa}P_{k_3,\kappa}Q^{\pm}_{<k_3+(\epsilon-1)i}\psi_3$. Thus we now replace both $P_{k_{1,3}}\psi_{1,3}$ by

$$
\sum_{\pm}\sum_{\kappa\in K_{\frac{\epsilon-1}{2}i}}\chi_{i,\mp\kappa}P_{k_{1,3}}Q^{\pm}_{<k_{1,3}+(\epsilon-1)i}\psi_{1,3},
$$

respectively. Observe that on account of the rapid decay properties of the kernels of the multipliers $Q_{<>-i\delta}$ etc., we can rewrite our term (up to errors of order of magnitude $2^{-Ni}$) as

$$
\begin{aligned}
&\sum_{\pm,\pm}\ \sum_{\kappa_{1,2}\in K_{\frac{\epsilon-1}{2}i},\mathrm{dist}(\pm\kappa_1,\pm\kappa_2)\leq C2^{\frac{\epsilon-1}{2}i}} P_{k_0,\kappa_2}Q^{\pm}_{<-i\delta}\chi_i(t)[\chi_{i,\mp\kappa_1}P_{k_1,\kappa_1}Q^{\pm}_{<(\epsilon-1)i+k_1}\psi_1\\
&\qquad\sum_{j=1,2}\triangle^{-1}\partial_j\Box IP_kQ_{>-i\delta+C}[\nabla^{-1}P_{k_2}Q_{<-\delta i}\delta\psi_2\chi_{i,\mp\kappa_2}P_{k_3,\kappa_2}Q^{\pm}_{<k_3+(\epsilon-1)i}\psi_3]]
\end{aligned}
$$

We can now invoke (2.4) as well as (3.7) in order to conclude that the preceding expression is bounded with respect to $||.||_{N[k_0]}$ by

$$
\begin{aligned}
&\leq C2^{\delta i}2^{\frac{\min\{k_i,k\}}{2}}(\sum_{\pm}\sum_{\kappa_1\in K_{\frac{\epsilon-1}{2}i}}||\chi_{i,\mp\kappa_1}P_{k_1,\kappa_1}Q^{\pm}_{<(\epsilon-1)i+k_1}\psi_1||^2_{PW[\pm\kappa_1]})^{\frac{1}{2}}\\
&\qquad(\sum_{\pm}\sum_{\kappa_3\in K_{\frac{\epsilon-1}{2}i}}||\chi_{i,\mp\kappa_3}P_{k_1,\kappa_3}Q^{\pm}_{<(\epsilon-1)i+k_1}\psi_1||^2_{PW[\pm\kappa_3]})^{\frac{1}{2}}||P_{k_2}\delta\psi_2||_{S[k_2]}
\end{aligned}
$$

On the other hand, it is easily seen that

$$(\sum_{\kappa_1\in K_{\frac{\epsilon-1}{2}i}}||\chi_{i,\mp\kappa_1}P_{k_1,\kappa_1}Q^{\pm}_{<(\epsilon-1)i+k_1}\psi_1||^2_{PW[\pm\kappa_1]})^{\frac{1}{2}}\leq C2^{\frac{\epsilon-1}{2}i}||P_{k_1}\psi_1||_{A[k_1]+\dot{X}^{0,\frac{1}{2},1}_{k_1}}.$$

This implies that the preceding expression may be bounded by $\leq C2^{-\mu i}\tilde{c}_{k_0}$, as desired. This concludes case (III).

(IV): *The term* **(IV)**. This is similar to (II) and the preceding case (III). Details are omitted. We are now done with establishing Proposition 2.2 for the trilinear null-forms linear in the perturbation $\delta\psi$.

**(C): Estimating the trilinear null-forms at least quadratic in $\delta\psi$.** The claim here follows directly from 3.4(c) in conjunction with lemma 2.6, provided any function $P_k\psi_\nu$ present is of first type. The other case is treated just like for the expressions linear in $\delta\psi$. This completes the trilinear estimates.

**3.0.9. The quintilinear and higher order terms linear in the perturbation.** These turn out to be fairly simple to estimate on account of the favorable Strichartz type estimates available for the radial components $\psi_\nu$. We recall that these terms have the following schematic structure:

$$\mathbf{A}(\delta\psi_\nu,\psi_\nu)=\nabla_{x,t}[\delta\psi\nabla^{-1}[R_\nu\psi\nabla^{-1}[\psi\nabla^{-1}(\psi^2)]]]$$
$$\mathbf{B}(\delta\psi_\nu,\psi_\nu)=\nabla_{x,t}[\psi\nabla^{-1}[R_\nu\delta\psi\nabla^{-1}[\psi\nabla^{-1}(\psi^2)]]]$$
$$\mathbf{C}(\delta\psi_\nu,\psi_\nu)=\nabla_{x,t}[\psi\nabla^{-1}[R_\nu\psi\nabla^{-1}[\delta\psi\nabla^{-1}(\psi^2)]]]$$
$$\mathbf{D}(\delta\psi_\nu,\psi_\nu)=\nabla_{x,t}[\psi\nabla^{-1}[R_\nu\psi\nabla^{-1}[\psi\nabla^{-1}(\delta\psi\psi)]]]$$
$$\mathbf{E}(\delta\psi_\nu,\psi_\nu)=\nabla_{x,t}[\nabla^{-1}[\psi\nabla^{-1}(\psi^2)]\nabla^{-1}IQ_{\nu j}(\delta\psi,\psi)]\text{ etc}$$
$$\mathbf{F}(\delta\psi_\nu,\psi_\nu)=\nabla_{x,t}[\delta\psi\nabla^{-1}[\nabla^{-1}(\psi\nabla^{-1}(\psi^2))\nabla^{-1}[\psi\nabla^{-1}(\psi^2)]]]\text{ etc}$$

We have left out the other terms like **E**, **F**, in which $\delta\psi$ gets shifted to different positions. We treat here the first quintilinear term in the list, the other terms being tedious reiterations of similar computations.

**(A)**: We microlocalize this term as follows:

$$\nabla_{x,t}P_{k_0}[P_{k_1}\delta\psi\nabla^{-1}P_{k_2}[P_{k_3}R_\nu\psi\nabla^{-1}P_{k_4}[P_{k_5}\psi\nabla^{-1}P_{k_6}(P_{k_7}\psi P_{k_8}\psi)]]]$$

We first dispose of the case when $\nu=0$ $R_\nu\psi$ has elliptic microsupport[17], i. e. is replaced by $R_0P_{k_3}Q_{\geq k_3+100}\psi$. Recall that the expression

$$\nabla^{-1}P_{k_4}[P_{k_5}\psi\nabla^{-1}P_{k_6}(P_{k_7}\psi P_{k_8}\psi)]$$

arises upon substituting the elliptic part for an input of the form $P_{k_4}\psi_j$, where $j\neq 0$. We then rewrite this expression as $P_{k_4}\psi_j-R_j\sum_{i=1,2}R_i\psi_i$, and are led to a term of the schematic form

$$\nabla_{x,t}P_{k_0}[P_{k_1}\delta\psi\nabla^{-1}P_{k_2}[P_{k_3}Q_{\geq k_3+100}R_0\psi P_{k_4}\psi]]$$

We claim the estimate (under our bootstrap assumption)

$$\nabla_{x,t}P_{k_0}[P_{k_1}\delta\psi\nabla^{-1}P_{k_2}[P_{k_3}Q_{\geq k_3+100}R_0\psi P_{k_4}\psi]]$$
$$\leq C2^{-\delta_1|k_1-k_0|}2^{\delta_2[\min_{i=2,3,4}\{k_i\}-\max_{i=2,3,4}\{k_i\}]}\frac{\tilde{c}_{k_4}}{\epsilon}\tilde{c}_{k_0}$$

[17]When $\nu\neq 0$, this distinction is irrelevant.

As usual, we may assume $k_0 = 0$. We first treat the case when $k_2 < -10$. Then consider the cases

**(a)**: $k_2 \in [k_3 - 10, k_3 + 10]$ whence $k_4 \leq k_3 + O(1)$. Either $P_{k_4}\psi$ has modulation at least comparable to that of $P_{k_3}Q_{\geq k_3}\psi$, or else either $P_{k_1}\delta\psi$ or the output has modulation at least comparable. These are all similar, so we treat the first possibility. Write this contribution as

$$\sum_{a \geq k_3+100} \nabla_{x,t}P_{k_0}[P_{k_1}\delta\psi\nabla^{-1}P_{k_2}[P_{k_3}Q_aR_0\psi P_{k_4}Q_{\geq a+O(1)}\psi]]$$

Then we can estimate

$$\begin{aligned}
&\|\sum_{a \geq k_3+100} \nabla_{x,t}P_{k_0}Q_{<k_0}[P_{k_1}\delta\psi\nabla^{-1}P_{k_2}[P_{k_3}Q_aR_0\psi P_{k_4}Q_{\geq a+O(1)}\psi]]\|_{L_t^1\dot{H}^{-1}} \\
&\leq C\sum_{a \geq k_3+100} \|P_{k_1}\delta\psi\|_{L_t^\infty L_x^2}\|P_{k_3}Q_aR_0\psi\|_{L_t^2L_x^2}\|P_{k_4}Q_{\geq a+O(1)}\psi\|_{L_t^2L_x^\infty} \\
&\leq C\sum_{a \geq k_3+100} 2^{-\frac{k_3}{2}}2^{(1-\epsilon)(k_4-a)}2^{\frac{k_4}{2}}\tilde{c}_{k_1}\frac{\tilde{c}_{k_3}}{\epsilon}\frac{\tilde{c}_{k_4}}{\epsilon},
\end{aligned}$$

irrespective of whether $P_{k_{3,4}}\psi$ is of the first or 2nd type. This is an acceptable bound. Similarly, we have

$$\begin{aligned}
&\|\sum_{a \geq k_3+100} \nabla_{x,t}P_{k_0}Q_{\geq k_0}[P_{k_1}\delta\psi\nabla^{-1}P_{k_2}[P_{k_3}Q_aR_0\psi P_{k_4}Q_{\geq a+O(1)}\psi]]\|_{\dot{X}_{k_0}^{-\frac{1}{2},-1,2}} \\
&\leq C\sum_{a \geq k_3+100} 2^{-\frac{k_0}{2}}\|P_{k_1}\delta\psi\|_{L_t^\infty L_x^2}\|P_{k_3}Q_aR_0\psi\|_{L_t^\infty L_x^2}\|P_{k_4}Q_{\geq a+O(1)}\psi\|_{L_t^2L_x^\infty} \\
&\leq C\sum_{a \geq k_3+100} 2^{\frac{k_4-k_0}{2}}2^{(1-\epsilon)(k_4-a)}\frac{\tilde{c}_{k_4}}{\epsilon}\tilde{c}_{k_0},
\end{aligned}$$

again an acceptable bound.

**(b)**: $k_2 < k_3 - 10$, whence $k_4 = k_3 + O(1)$. One proceeds as in the preceding case. Again treating the case when $P_{k_4}\psi_4$ is at modulation at least comparable to that of $P_{k_3}\psi_3$, we can estimate

$$\begin{aligned}
&\|\sum_{a \geq k_3+100} \nabla_{x,t}P_{k_0}Q_{<k_0}[P_{k_1}\delta\psi\nabla^{-1}P_{k_2}[P_{k_3}Q_aR_0\psi P_{k_4}Q_{\geq a+O(1)}\psi]]\|_{L_t^1\dot{H}^{-1}} \\
&2^{k_2}\|P_{k_1}\delta\psi\|_{L_t^\infty L_x^2}\|P_{k_3}Q_aR_0\psi\|_{L_t^2L_x^2}\|P_{k_4}Q_{\geq a+O(1)}\psi\|_{L_t^2L_x^2} \\
&\leq C\sum_{a \geq k_3+100} 2^{k_2-k_3}2^{(1-\epsilon)(k_3-a)}\frac{\tilde{c}_{k_3}}{\epsilon}\frac{\tilde{c}_{k_4}}{\epsilon}\tilde{c}_{k_0},
\end{aligned}$$

which leads to an acceptable estimate. The contribution of $P_{k_0}Q_{\geq k_0}$ is treated similarly.

**(c)**: The case $k_3 \leq k_2 - 10$. Rewrite the term as

$$\begin{aligned}
&\nabla_{x,t}P_{k_0}[P_{k_1}\delta\psi_1\nabla^{-1}P_{k_4}\psi P_{k_3}R_0(1-I)\psi] = \\
&\qquad\qquad\sum_{a > k_3+100} \nabla_{x,t}P_{k_0}[P_{k_1}\delta\psi_1\nabla^{-1}P_{k_4}\psi P_{k_3}R_0Q_a\psi]
\end{aligned}$$

Then one estimates

$$\begin{aligned}
&||\nabla_{x,t}P_{k_0}Q_{<k_0}[Q_{\geq a-10}[P_{k_1}\delta\psi_1\nabla^{-1}P_{k_4}\psi]P_{k_3}R_0Q_a\psi]||_{L_t^1\dot{H}^{-1}}\\
&\leq C||Q_{\geq a-10}[P_{k_1}\delta\psi_1\nabla^{-1}P_{k_4}\psi]||_{L_t^2L_x^2}||P_{k_3}R_0Q_a\psi||_{L_t^2L_x^\infty}\\
&\leq C2^{-\frac{a}{2}}2^{\frac{\min\{a-k_4,0\}}{4+}}||P_{k_1}\delta\psi_1||_{S[k_1]}||P_{k_4}\psi_4||_{\dot{X}_{k_4}^{0,\frac{1}{2},1}+A[k_4]}\\
&\qquad 2^{\frac{k_3}{2}}2^{\mu(a-k_3)}||P_{k_3}R_0Q_a\psi||_{\dot{X}_{k_4}^{-(\frac{1}{2}-\mu),1-\mu,1}}\\
&\leq C2^{\min\{\frac{a-k_4}{4+},0\}}2^{\frac{k_3-a}{2+}}\tilde{c}_{k_1}\frac{\tilde{c}_{k_4}}{\epsilon}
\end{aligned}$$

One can sum over $a > k_3 + 100$ to obtain the desired estimate. Similarly, we have

$$\begin{aligned}
&||\nabla_{x,t}P_{k_0}Q_{<k_0}[Q_{<a-10}[P_{k_1}\delta\psi_1\nabla^{-1}P_{k_4}\psi]P_{k_3}R_0Q_a\psi]||_{\dot{X}_{k_0}^{-1,-\frac{1}{2},1}}\\
&\leq C2^{-\frac{a}{2}}||Q_{<a-10}[P_{k_1}\delta\psi_1\nabla^{-1}P_{k_4}\psi]||_{L_t^\infty L_x^2}||P_{k_3}R_0Q_a\psi||_{L_t^2L_x^\infty}\\
&\leq C2^{\min\{\frac{a-k_4}{4+},0\}}2^{\frac{k_3-a}{2+}}\frac{\tilde{c}_{k_3}}{\epsilon}\tilde{c}_{k_1}
\end{aligned}$$

Next, as to the large modulation contribution of the output, we estimate using theorem 2.3

$$\begin{aligned}
&||\nabla_{x,t}P_{k_0}Q_{\geq k_0}[P_{k_1}\delta\psi_1\nabla^{-1}P_{k_4}\psi P_{k_3}R_0Q_a\psi]||_{\dot{X}_{k_0}^{-\frac{1}{2},-1,2}}\\
&\leq C2^{-\frac{k_0}{2}}||P_{k_1}\delta\psi_1||_{L_t^\infty L_x^2}||\nabla^{-1}P_{k_4}\psi P_{k_3}R_0Q_a\psi||_{L_t^2L_x^\infty}\\
&\leq C2^{\delta(k_3-k_4)}2^{\frac{k_4-k_0}{2}}\tilde{c}_{k_1}\frac{\tilde{c}_{k_4}}{\epsilon}
\end{aligned}$$

Now consider the case $k_2 > 10$, whence $k_1 = k_2 + O(1)$. Again this only requires trivial modifications: for example, one estimates in case $k_3 = k_4 + O(1)$

$$\begin{aligned}
&||\sum_{a\geq k_3+100}\nabla_{x,t}P_{k_0}Q_{<k_0}[P_{k_1}\delta\psi\nabla^{-1}P_{k_2}[P_{k_3}Q_aR_0\psi P_{k_4}Q_{\geq a+O(1)}\psi]]||_{L_t^1\dot{H}^{-1}}\\
&\leq C\sum_{a\geq k_3+100}2^{-k_2}||P_{k_1}\delta\psi||_{L_t^\infty L_x^2}||P_{k_3}Q_aR_0\psi||_{L_t^2L_x^2}||P_{k_4}Q_{\geq a+O(1)}\psi||_{L_t^2L_x^2}\\
&\leq C\sum_{a\geq k_3+100}2^{-k_3}2^{(1-\epsilon)(k_3-a)}\tilde{c}_{k_1}\frac{\tilde{c}_{k_4}}{\epsilon}\leq C2^{-(1-\epsilon)k_1}2^{(1-\epsilon)(k_1-k_3)}\tilde{c}_{k_0}\frac{\tilde{c}_{k_4}}{\epsilon},
\end{aligned}$$

and this is again an acceptable estimate. The remaining cases are more of the same. We conclude from this[18] that the estimate

$$||\chi_i(t)P_{k_0}[P_{k_1}\delta\psi\nabla^{-1}P_{k_2}[P_{k_3}(1-I)R_0\psi P_{k_4}\psi]]||_{N[k_0]}\leq C2^{-\mu i}\tilde{c}_{k_0}$$

holds, provided we have $i \leq C\max_{i=2,3,4}\{|k_i|\}$. Now assume that $\max\{|k_{2,3,4}|\} << i$. We shall treat this quantity as $O(1)$. Also, assume $i \leq Ck_1$. We also omit the operator $\nabla^{-1}P_{k_2}$ for simplicity's sake, and estimate

$$\begin{aligned}
&||\nabla_{x,t}P_{k_0}Q_{>k_0}[P_{k_1}\delta\psi P_{k_4}\psi R_0(1-I)\psi||_{N[k_0]}\\
&\leq C2^{-\frac{k_0}{2}}||P_{k_1}\delta\psi||_{L_t^\infty L_x^2}||P_{k_4}\psi R_0(1-I)\psi||_{L_t^2L_x^2}\leq C2^{-\frac{k_0}{2}}2^{\frac{k_4}{2}}\tilde{c}_{k_1}\frac{\tilde{c}_{k_4}}{\epsilon}
\end{aligned}$$

[18]Applying the multiplier $\chi_i(t)$ doesn't affect the estimate for reasons explained earlier.

This yields the desired exponential gain in $i$. The contribution of the hyperbolic part is unfortunately a bit more complicated. First, observe that if $a > \delta i$, one estimates

$$||\nabla_{x,t} P_{k_0} Q_{\leq k_0}[Q_{<a-10}[P_{k_1}\delta\psi P_{k_4}\psi]R_0 Q_a\psi||_{N[k_0]}$$
$$\leq C2^{-\frac{a}{2}}||P_{k_1}\delta\psi||_{L_t^\infty L_x^2}||P_{k_4}\psi||_{L_t^\infty L_x^\infty}||R_0 Q_a P_{k_3}\psi||_{L_t^2 L_x^\infty} \leq C2^{-\frac{a}{2+}}\tilde{c}_{k_1}\frac{\tilde{c}_{k_4}}{\epsilon},$$

which upon summing over $a > \delta i$ results in the desired exponential gain in $i$. The case when $Q_{<a-10}$ is replaced by $Q_{\geq a-10}$ is similar (place the output into $L_t^1\dot{H}^{-1}$). Now consider

$$\nabla_{x,t} P_{k_0} Q_{\leq k_0}\chi_i(t)[[P_{k_1}\delta\psi P_{k_4}\psi]R_0(1-I)P_{k_3}Q_{<\delta i}\psi]$$

We have

$$||\nabla_{x,t} P_{k_0} Q_{\leq k_0}\chi_i(t)[Q_{<a-10}[P_{k_1}\delta\psi P_{k_4}\psi]R_0(1-I)P_{k_3}Q_a\psi]||_{N[k_0]}$$
$$\leq C2^{-\frac{a}{2}}||P_{k_1}\delta\psi||_{L_t^\infty L_x^2}||\chi_i(t)P_{k_4}\psi_4||_{L_t^\infty L_x^\infty}||R_0(1-I)P_{k_3}Q_a\psi||_{L_t^2 L_x^2}$$
$$\leq C2^{-\frac{i}{2+}}\tilde{c}_{k_1}\frac{\tilde{c}_{k_3}}{\epsilon}$$

Summing over $a < \delta i$ yields an acceptable estimate. Now consider for $k_3 + 100 < a < \delta i$

$$\nabla_{x,t} P_{k_0} Q_{\leq k_0}\chi_i(t)[Q_{\geq a-10}[P_{k_1}\delta\psi P_{k_4}\psi]R_0(1-I)P_{k_3}Q_a\psi]$$

We first reduce $P_{k_1}\delta\psi_1$ to modulation $< 2^{a-100}$, which is straightforward. We estimate

$$||\nabla_{x,t} P_{k_0} Q_{\leq k_0}[Q_{\geq a-10}[P_{k_1}Q_{<a-100}\delta\psi$$
$$P_{k_4}Q_{>k+(\epsilon-1)i}(\chi_i(t)\psi)]R_0(1-I)P_{k_3}Q_a\psi]||_{L_t^1\dot{H}^{-1}}$$
$$\leq C2^{-\frac{a}{2}}||P_{k_1}Q_{<a-100}\delta\psi P_{k_4}Q_{>k+(\epsilon-1)i}(\chi_i(t)\psi)]||_{\dot{X}_{k_1}^{0,\frac{1}{2},\infty}}||R_0(1-I)P_{k_3}Q_a\psi||_{L_t^2 L_x^2}$$
$$\leq C2^{-\frac{a}{2}}||P_{k_1}Q_{<a-100}\delta\psi||_{S[k_1]}||P_{k_4}Q_{>k+(\epsilon-1)i}(\chi_i(t)\psi)||_{\dot{X}_{k_4}^{0,\frac{1}{2},1}}||R_0(1-I)P_{k_3}Q_a\psi||_{L_t^2 L_x^2}$$
$$\leq C2^{-\frac{a}{2+}}2^{-\mu i}\tilde{c}_{k_1},$$

using lemma 3.3. Next, borrowing notation form the proof of lemma 3.3, estimate

$$||\nabla_{x,t} P_{k_0} Q_{\leq k_0}[Q_{\geq a-10}[P_{k_1}Q_{<a-100}\delta\psi\phi_i(t,x)$$
$$P_{k_4}Q_{<k+(\epsilon-1)i}(\chi_i(t)\psi)]R_0(1-I)P_{k_3}Q_a\psi]||_{L_t^1\dot{H}^{-1}}$$
$$\leq C||P_{k_1}Q_{<a-100}\delta\psi||_{L_t^\infty L_x^2}||\phi_i(t,x)P_{k_4}\psi||_{L_t^2 L_x^\infty}||R_0(1-I)P_{k_3}Q_a\psi||_{L_t^2 L_x^2}$$
$$+ ||Q_{\geq a-10}[P_{k_1}Q_{<a-100}\delta\psi\phi_i(t,x)Q_{>k+(\epsilon-1)i}\psi]||_{L_t^2 L_x^2}||R_0(1-I)P_{k_3}Q_a\psi||_{L_t^2 L_x^2}$$

The 2nd summand may be estimated as in the immediately preceding since we may assume that $|a|, |k|$ etc are $<< \epsilon i$. As to the first summand, we estimate it using

$$||\phi_i(t)P_{k_4}\psi||_{L_t^2 L_x^\infty} \leq C2^{-\mu i},$$

from which the desired estimate follows. Now one decomposes

(3.8)
$$(1-\phi_i(t,x))P_{k_4}Q_{<k+(\epsilon-1)i}(\chi_i(t)\psi) = \sum_{\pm}\sum_{\kappa\in K_{\frac{\epsilon-1}{2}i}}\chi_{i,\mp\kappa}^c P_{k_4,\kappa}Q_{<k+(\epsilon-1)i}^{\pm}(\chi_i(t)\psi)$$
$$+\sum_{\pm}\sum_{\kappa\in K_{\frac{\epsilon-1}{2}i}}\chi_{i,\mp\kappa}P_{k_4,\kappa}Q_{<k+(\epsilon-1)i}^{\pm}(\chi_i(t)\psi),$$

and proceeds as in the trilinear estimates: plugging in the first summand on the right, we have

$$||Q_{\geq a-10}[P_{k_1}Q_{<a-100}\delta\psi\sum_{\pm}\sum_{\kappa\in K_{\frac{\epsilon-1}{2}i}}\chi^c_{i,\mp\kappa}P_{k_4,\kappa}Q^{\pm}_{<k+(\epsilon-1)i}(\chi_i(t)\psi)]||_{\dot{X}^{0,\frac{1}{2},\infty}_{k_1}}$$
$$\leq C(\sum_{\tilde{\kappa}_1\in K_{\frac{a-k_4}{2}}}||P_{k_1,\tilde{\kappa}_1}Q^{\pm}_{<a-100}\delta\psi||^2_{NFA^*[\pm\kappa_1]})^{\frac{1}{2}}$$
$$(\sum_{\tilde{\kappa}_2\in K_{\frac{a-k_4}{2}}}||P_{k_4,\tilde{\kappa}_4}\sum_{\kappa\in K_{\frac{\epsilon-1}{2}i}}\chi^c_{i,\mp\kappa}P_{k_4,\kappa}Q^{\pm}_{<k+(\epsilon-1)i}(\chi_i(t)\psi)||^2_{PW[\pm\tilde{\kappa}_4]})^{\frac{1}{2}}$$

This can be estimated by $\leq C2^{-\mu i}\tilde{c}_{k_1}$, as desired. Plugging in the 2nd part of (3.8) is handled as in the trilinear estimates, exploiting the orthogonality of these pieces. This gives the desired estimate. The case when $k_1 << i$ is more elementary. One may always assume that $R_0\psi_3$ lives at modulation $< \delta i$ (argue as before), whence one may estimate

$$||\nabla_{x,t}P_{k_0}Q_{\leq k_0}\chi_i(t)[[P_{k_1}\delta\psi P_{k_4}\psi]R_0(1-I)P_{k_3}Q_{<\delta i}\psi]||_{L^1_t\dot{H}^{-1}}$$
$$\leq C||P_{k_1}\delta\psi||_{L^M_tL^{2+}_x}||P_{k_4}\chi_i(t)\psi||_{L^{2+}_tL^\infty_x}||R_0(1-I)P_{k_3}Q_{<\delta i}\psi||_{L^2_tL^M_x}$$

Choosing $\frac{1}{2}-\frac{1}{2+} >> \delta$ results in the desired estimate $\leq C2^{-\mu i}\tilde{c}_{k_1}$. Also, we have

$$||\nabla_{x,t}P_{k_0}Q_{>k_0}\chi_i(t)[[P_{k_1}\delta\psi P_{k_4}\psi]R_0(1-I)P_{k_3}Q_{<\delta i}\psi]||_{\dot{X}^{-\frac{1}{2},-1,2}_{k_0}}$$
$$\leq C||P_{k_1}\delta\psi||_{L^\infty_tL^2_x}||\chi_i(t)P_{k_4}\psi||_{L^\infty_tL^\infty_x}||R_0(1-I)P_{k_3}Q_{<\delta i}\psi||_{L^2_tL^2_x}\leq C2^{-\frac{i}{2+}}\tilde{c}_{k_1}$$

which is as desired. This finally concludes estimating the contribution when $P_{k_3}R_\nu\psi$ is replaced by $P_{k_3}(1-I)R_\nu\psi$. Thus we may replace $P_{k_3}R_\nu\psi$ by $P_{k_3}IR_\nu\psi$, whence we may discard $IR_\nu$ for all intents and purposes. We revert to the original formulation of this term given by $\mathbf{A}(\delta\psi,\psi)$, and decompose the innermost bracket $(\psi^2)$ into a $Q_{\nu j}$-type null-form as well as error terms at least quadrilinear by means of our standard Hodge-type decompositions, i. e. we write (using schematic notation)

$$(\psi^2)=R_\nu\psi^1R_j\psi^2-R_j\psi^1R_\nu\psi^2+\nabla^{-1}(\psi\nabla^{-1}(\psi^2))R_\nu\psi$$
$$+\nabla^{-1}(\psi\nabla^{-1}(\psi^2))\nabla^{-1}(\psi\nabla^{-1}(\psi^2)).$$

Substituting the $Q_{\nu j}$-form for now, we claim that we have the estimate

$$||\nabla_{x,t}P_{k_0}[P_{k_1}\delta\psi\nabla^{-1}P_{k_2}[P_{k_3}\psi\nabla^{-1}P_{k_4}[P_{k_5}\psi\nabla^{-1}P_{k_6}Q_{\nu j}(P_{k_7}\psi P_{k_8}\psi)]]]||_{N[k_0]}$$
$$\leq C\tilde{c}_{k_1}\tilde{c}_{k_3}\tilde{c}_{k_5}(\tilde{c}_{k_7}+\tilde{c}_{k_8})$$
$$2^{\delta_1[\min\{k_7,k_8\}-\max\{k_7,k_8\}]}2^{\delta_2[\min_{i=2,\ldots,6}\{k_i\}-\max_{i=2,\ldots,6}]\{k_i\}}$$

To prove this, one needs to analyze the possible frequency interactions. By scaling invariance, we may assume $k_0=0$. The first step consists in reducing $P_{k_6}Q_{\nu j}(P_{k_7}\psi P_{k_8}\psi)$ to hyperbolic microsupport, dealing with the contribution of $(1-I)P_{k_6}Q_{\nu j}(P_{k_7}\psi P_{k_8}\psi)$. The argument for this is given in the appendix[19] of [**23**]. Then we deal with the following cases in schematic fashion:

(**A**.a): $k_2<-10$, $k_5=k_6+O(1)$, $k_3=k_4+O(1)$. We combine 3.4(b) as well as lemma 3.2 to conclude that

$$||P_{k_6}Q_{\nu j}I(P_{k_7}\psi P_{k_8}\psi)||_{L^2_tL^2_x}\leq C2^{\frac{\min\{k_{6,7,8}\}}{2}}(\tilde{c}_{k_7}+\tilde{c}_{k_8}).$$

[19]If one of the inputs of $Q_{\nu j}(.)$ is of 2nd type, one use lemma 3.2 instead of 3.4(b).

Next, we have

$$||P_{k_4}[P_{k_5}\psi\nabla^{-1}P_{k_6}A]||_{L_t^{\frac{4}{3}}L_x^2} \leq C2^{-k_6}(\sum_{c\in C_{k_5,k_4-k_5}} ||P_c\psi||^2_{L_t^4L_x^\infty})^{\frac{1}{2}}||P_{k_6}A||_{L_t^2L_x^2}$$

$$\leq C2^{\frac{3k_5}{4}}2^{\frac{k_4-k_5}{2}}2^{-k_6}||P_{k_5}\psi||_{S[k_5]}||P_{k_6}A||_{L_t^2L_x^2}$$

Finally, we have

$$||P_{k_2}[P_{k_3}\psi\nabla^{-1}P_{k_4}A]||_{L_t^1L_x^2} \leq C2^{-k_4}(\sum_{c\in C_{k_3,k_2-k_3}} ||P_c\psi||^2_{L_t^4L_x^\infty})^{\frac{1}{2}}||P_{k_4}A||_{L_t^{\frac{4}{3}}L_x^2}$$

$$\leq C2^{\frac{3k_3}{4}}2^{\frac{k_2-k_3}{2}}2^{-k_4}||P_{k_3}\psi||_{S[k_3]}||P_{k_4}A||_{L_t^{\frac{4}{3}}L_x^2}$$

We have used here that $(\sum_{c\in C_{k_4,k_2-k_4}}||P_c\psi_4||^2_{L_t^{\frac{4}{3}}L_x^2})^{\frac{1}{2}} \leq C||P_{k_4}\psi_4||_{L_t^{\frac{4}{3}}L_x^2}$. Combining these inequalities, we easily see that

$$||\nabla_{x,t}P_{k_0}Q_{<k_0}[P_{k_1}\delta\psi\nabla^{-1}P_{k_2}[P_{k_3}\psi\nabla^{-1}P_{k_4}[P_{k_5}\psi\nabla^{-1}P_{k_6}IQ_{\nu j}(P_{k_7}\psi P_{k_8}\psi)]]]]||_{L_t^1\dot{H}^{-1}}$$
$$\leq C||P_{k_1}\delta\psi||_{L_t^\infty L_x^2}||P_{k_2}[P_{k_3}\psi\nabla^{-1}P_{k_4}[P_{k_5}\psi\nabla^{-1}P_{k_6}Q_{\nu j}(P_{k_7}\psi P_{k_8}\psi)]]]||_{L_t^1L_x^2}$$
$$\leq C2^{\frac{3k_3}{4}+\frac{k_2-k_3}{2}-k_4+\frac{3k_5}{4}+\frac{k_4-k_5}{2}+\frac{\min\{k_{6,7,8}\}}{2}-k_6}$$
$$||P_{k_1}\delta\psi||_{L_t^\infty L_x^2}||P_{k_3}\psi||_{S[k_3]}||P_{k_5}\psi||_{S[k_5]}(\tilde{c}_{k_7}+\tilde{c}_{k_8}).$$

Moreover, we can estimate

$$||\nabla_{x,t}P_{k_0}Q_{\geq k_0}[P_{k_1}\delta\psi\nabla^{-1}P_{k_2}[P_{k_3}\psi\nabla^{-1}P_{k_4}[P_{k_5}\psi\nabla^{-1}P_{k_6}IQ_{\nu j}(P_{k_7}\psi P_{k_8}\psi)]]]]||_{\dot{X}_{k_0}^{-\frac{1}{2},-1,2}}$$
$$\leq C2^{-\frac{k_0}{2}}2^{k_2}||P_{k_1}\delta\psi||_{L_t^\infty L_x^2}||P_{k_2}[P_{k_3}\psi\nabla^{-1}P_{k_4}[P_{k_5}\psi\nabla^{-1}P_{k_6}IQ_{\nu j}(P_{k_7}\psi P_{k_8}\psi)]]]||_{L_t^2L_x^1}$$
$$\leq C2^{k_2-\frac{k_0}{2}}2^{\frac{\min\{k_{6,7,8}\}}{2}-k_6}||P_{k_1}\delta\psi||_{L_t^\infty L_x^2}||P_{k_3}\psi_3||_{L_t^\infty L_x^2}||P_{k_5}\psi_5||_{L_t^\infty L_x^2}(\tilde{c}_{k_7}+\tilde{c}_{k_8}),$$

One easily verifies that these estimates verify the claim in the case under consideration.

(**A**.b): $k_2+O(1) \leq k_6 << k_5$. In this case, we rewrite the term under consideration as

$$\nabla_{x,t}P_{k_0}[P_{k_1}\delta\psi\nabla^{-1}P_{k_2}[(P_{k_3}\psi\nabla^{-1}P_{k_5}\psi)\nabla^{-1}P_{k_6}IQ_{\nu j}(P_{k_7}\psi P_{k_8}\psi)]]].$$

Note that necessarily $k_3 = k_5 + O(1)$. We have

$$||\nabla_{x,t}P_{k_0}Q_{<k_0}[P_{k_1}\delta\psi\nabla^{-1}P_{k_2}[(P_{k_3}\psi\nabla^{-1}P_{k_5}\psi)\nabla^{-1}P_{k_6}IQ_{\nu j}(P_{k_7}\psi P_{k_8}\psi)]]]||_{L_t^1\dot{H}^{-1}}$$
$$\leq C2^{\frac{4k_2}{M}-k_6}||P_{k_1}\delta\psi||_{L_t^\infty L_x^2}||P_{\leq k_6+O(1)}(P_{k_3}\psi\nabla^{-1}P_{k_5}\psi)||_{L_t^2L_x^{\frac{M}{2}}}$$
$$||P_{k_6}IQ_{\nu j}(P_{k_7}\psi P_{k_8}\psi)||_{L_t^2L_x^2}$$

Then we estimate

$$||P_{\leq k_6+O(1)}(P_{k_3}\psi\nabla^{-1}P_{k_5}\psi)||_{L_t^2L_x^{\frac{M}{2}}} \leq C\sum_{a<k_6+O(1)}||P_a(P_{k_3}\psi\nabla^{-1}P_{k_5}\psi)||_{L_t^2L_x^{\frac{M}{2}}}$$
$$\leq C\sum_{a<k_6+O(1)}(\sum_{c\in C_{k_3,a-k_3}}||P_c\psi||^2_{L_t^4L_x^M})^{\frac{1}{2}}(\sum_{c\in C_{k_5,a-k_5}}||P_c\psi||^2_{L_t^4L_x^M})^{\frac{1}{2}}$$
$$\leq C\sum_{a<k_6+O(1)}2^{\frac{a-k_3}{2+}}2^{\frac{a-k_5}{2+}}2^{(\frac{3}{4}-\frac{2}{M})(k_3+k_5)}2^{-k_3}||P_{k_3}\psi||_{S[k_3]}||P_{k_5}\psi||_{S[k_5]},$$

where $2+ = 2 + (M)$. Putting this together with the above estimate for $IQ_{\nu j}(P_{k_7}\psi P_{k_8}\psi)$ easily results in the claim for this case as well. Replacing the operator $Q_{<k_0}$ by $Q_{\geq k_0}$, we have

$$||\nabla_{x,t} P_{k_0} Q_{\geq k_0}[P_{k_1}\delta\psi\nabla^{-1}P_{k_2}[(P_{k_3}\psi\nabla^{-1}P_{k_5}\psi)\nabla^{-1}P_{k_6}IQ_{\nu j}(P_{k_7}\psi P_{k_8}\psi)]]||_{\dot{X}_{k_0}^{-\frac{1}{2},-1,2}}$$
$$\leq C2^{k_2}||P_{k_1}\delta\psi||_{L_t^\infty L_x^2}||P_{k_2}[(P_{k_3}\psi\nabla^{-1}P_{k_5}\psi)\nabla^{-1}P_{k_6}IQ_{\nu j}(P_{k_7}\psi P_{k_8}\psi)]||_{L_t^2 L_x^1}$$
$$\leq C2^{\frac{4(k_2-k_5)}{M}}2^{\frac{1}{2+}(k_6-k_5)}2^{\frac{\min\{k_{6,7,8}\}-k_6}{2}}$$
$$||P_{k_1}\delta\psi||_{L_t^\infty L_x^2}||P_{k_3}\psi_3||_{L_t^\infty L_x^2}||P_{k_5}\psi_5||_{L_t^\infty L_x^2}(\tilde{c}_{k_7}+\tilde{c}_{k_8})$$

This again verifies the claim.
(**A**.c): $k_6 \leq k_2 + O(1) << k_5$. Consider the case when the output is reduced to modulation $< 2^{k_0}$, the opposite case being treated similarly. We rewrite the term as above and estimate

$$||\nabla_{x,t} P_{k_0} Q_{<k_0}[P_{k_1}\delta\psi\nabla^{-1}P_{k_2}[(P_{k_3}\psi\nabla^{-1}P_{k_5}\psi)\nabla^{-1}P_{k_6}IQ_{\nu j}(P_{k_7}\psi P_{k_8}\psi)]]||_{L_t^1\dot{H}^{-1}}$$
$$\leq C2^{(\frac{2}{M}-1)k_2}||P_{k_1}\delta\psi||_{L_t^\infty L_x^2}||(P_{k_3}\psi\nabla^{-1}P_{k_5}\psi)||_{L_t^2 L_x^{\frac{M}{2}}}||\nabla^{-1}P_{k_6}IQ_{\nu j}(P_{k_7}\psi P_{k_8}\psi)||_{L_t^2 L_x^\infty}$$
$$\leq C2^{\frac{\min\{k_{7,8}\}-\max\{k_7,k_8\}}{2}}2^{\frac{k_6-k_2}{2}}2^{\frac{k_2-k_3}{2+}}\tilde{c}_{k_1}\tilde{c}_{k_3}\tilde{c}_{k_5}(\tilde{c}_{k_7}+\tilde{c}_{k_8})$$

This again yields the desired claim.
(**A**.d): $k_5 \leq k_6 + O(1)$. This case is treated similarly and left out. The remaining frequency interactions are also simple variations of this kind of reasoning and left out. This establishes the claim for term $(\mathbf{A})(\delta\psi,\psi)$. We immediately deduce that if we reduce the expression to dyadic time $\sim 2^i$, i. e. we apply a multiplier $\chi_i(t)$ in front, we may reduce almost all (logarithmic) frequencies to norm $<< i$, i. e. we may assume $|k_2|,\ldots,|k_8| << i$. We shall treat these frequencies as $O(1)$. In that case, though, we can argue rather simply: observe that

$$||\nabla_{x,t} P_{k_0} Q_{<k_0}\chi_i(t)[P_{k_1}\delta\psi\nabla^{-1}P_{k_2}[(P_{k_3}\psi\nabla^{-1}P_{k_5}\psi)\nabla^{-1}P_{k_6}IQ_{\nu j}(P_{k_7}\psi P_{k_8}\psi)]]]||_{L_t^1\dot{H}^{-1}}$$
$$\leq C||P_{k_1}\delta\psi||_{L_t^\infty L_x^2}||P_{k_3}\psi||_{L_t^{2+}L_x^\infty}||P_{k_5}\psi||_{L_t^M L_x^{2+}}||\nabla^{-1}P_{k_6}IQ_{\nu j}(P_{k_7}\psi P_{k_8}\psi)||_{L_t^2 L_x^2}$$
$$\leq C2^{-\mu i}\tilde{c}_{k_0}.$$

We have used 3.4(c) as well as lemma 3.2. Similarly, we can estimate

$$||\nabla_{x,t} P_{k_0} Q_{\geq k_0}\chi_i(t)[P_{k_1}\delta\psi\nabla^{-1}P_{k_2}[(P_{k_3}\psi\nabla^{-1}P_{k_5}\psi)$$
$$\nabla^{-1}P_{k_6}IQ_{\nu j}(P_{k_7}\psi P_{k_8}\psi)]]]||_{\dot{X}_{k_0}^{-\frac{1}{2},-1,2}}$$
$$\leq C2^{-\frac{k_0}{2}}||P_{k_1}\delta\psi||_{L_t^\infty L_x^2}||\chi_i(t)\psi_3||_{L_t^\infty L_x^\infty}||P_{k_5}\psi||_{L_t^\infty L_x^\infty}||P_{k_6}IQ_{\nu j}(P_{k_7}\psi P_{k_8}\psi)||_{L_t^2 L_x^2}$$
$$\leq C2^{-\mu i}\tilde{c}_{k_0}$$

We still need to consider the case when $(\psi^2)$ gets replaced by the error terms $R_\nu\psi\nabla^{-1}(\psi\nabla^{-1}(\psi^2))$ etc. But this is straightforward: first, if $\nu = 0$, one reduces $R_0\psi$ to $IR_0\psi$ arguing as before; the latter is morally equivalent to $\psi$. One then winds up with the following schematic expression:

$$\nabla_{x,t}[\delta\psi\nabla^{-1}[\psi\nabla^{-1}[\psi\nabla^{-1}[\psi\nabla^{-1}[\psi\nabla^{-1}[\psi^2]]]]]]$$

One can place two of the inputs $\psi$ into $L_t^{2+}L_x^{\infty}$ and two others into $L_t^{\infty}L_x^2$, $L_t^M L_x^{2+}$, respectively[20]. Details are tedious reiterations of previously given arguments. This concludes the estimates for term $\mathbf{A}(\delta\psi, \psi)$. The estimates for the remaining terms $\mathbf{B}(\delta\psi, \psi)$ etc. are more of the same. One invokes the weaker form of the improved Strichartz norms available for the spaces $S[k]$ available by 3.4(f), which is good enough when combined with the stronger version available for the $\psi_\nu$.

**3.0.10. The higher order error terms at least quadratic in $\delta\psi$. Completing the proof of Proposition 2.2.** We need to estimate expressions like $\mathbf{A}(\delta\psi, \psi)$ as in the preceding section in which at least one additional $\psi$ has been replaced by $\delta\psi$. Of course we no longer need to gain in time, but we need to obtain an estimate like in theorem 2.5 upon frequency localizing the expression and its inputs. This will eventually allow us to sum over all frequency interactions. The degree of difficulty of these terms varies depending on how many inputs $\delta\psi$ are present, due to the weaker nature of the estimates satisfied by these. However, we shall try to treat all of these terms in a uniform manner, as these estimates are not interesting in and of themselves. Proceeding as in [**23**], one introduces a null-structure into these terms by splitting all inputs into gradient and elliptic parts, just as before. Substituting the elliptic parts results in error terms of higher order, while the gradient parts contribute to the null-structure. Of course, as before one may apply this splitting to both the inputs $\delta\psi$ as well as $\psi$. Carrying out this process a finite number of times results in the decomposition described in theorem 2.5. This theorem then ensures that all the resulting terms upon frequency localizing will satisfy the desired estimates, provided all inputs $P_{k_i}\psi$ are of the first type. Our concern here is whether the same kind of estimates hold if they are of the 2nd type. Thus consider for example a typical quintilinear term of the form

$$\nabla_{x,t}[P_{k_1}\delta\psi_1 P_{k_2}\nabla^{-1}[P_{k_3}R_\nu\delta\psi\nabla^{-1}[P_{k_4}\psi P_{k_5}\psi]]]$$

The first step here consists in reducing $R_\nu\delta\psi$ to the 'hyperbolic version ' $R_\nu I\delta\psi$, arguing as in the preceding subsection. Next, one applies the usual Hodge-type decomposition to the innermost square bracket $[\psi^2]$, replacing this by the sum of a $Q_{\nu j}$-type null-form as well as error terms at least quadrilinear. One treats the resulting quintilinear null-form just as in [**23**], resulting in the desired estimate, provided both inputs $P_{k_{4,5}}\psi_{4,5}$ are of the first type. Now assume at least one is of the 2nd type. We first consider the expression

$$\nabla_{x,t}[P_{k_1}\delta\psi_1 P_{k_2}\nabla^{-1}[P_{k_3}R_\nu I\delta\psi\nabla^{-1}P_{k_4}[P_{k_5}\psi\nabla^{-1}(1-I)Q_{\nu j}[P_{k_6}\psi P_{k_7}\psi]]]$$

If both $P_{k_{6,7}}\psi$ are of the first type, one argues here as in the appendix of [**23**], resulting in the desired estimate. If at least one of these inputs is of the 2nd type, one substitutes lemma 3.2 instead of 3.4(b) in that same argument. Thus we can now replace $Q_{\nu j}(P_{k_6}\psi, P_{k_7}\psi)$ by $IQ_{\nu j}(P_{k_6}\psi, P_{k_7}\psi)$. If both $P_{k_{6,7}}\psi$ are of first type, one again argues as in [**23**] (these quintilinear null-forms are part of the expansion in theorem 2.5). If one of $P_{k_{6,7}}$ is of 2nd type, the estimate becomes quite simple due to the strong estimates available: observe that then

$$||P_{k_5}\nabla^{-1}Q_{\nu j}I[P_{k_6}\psi, P_{k_7}\psi]||_{L_t^1 L_x^{2+}} \leq C2^{(1-\delta_1)\max\{k_i\}}2^{\delta_2\min\{k_i\}}\tilde{c}_{k_6}\tilde{c}_{k_7}$$

---

[20] One avoids losses for high-high interactions by interpolating with the improved $L_t^4 L_x^{\infty}$-norms.

Repeating the usual frequency trichotomies, one gets from here that

$$||\nabla_{x,t}P_{k_0}Q_{<k_0}[P_{k_1}\delta\psi_1 P_{k_2}\nabla^{-1}[P_{k_3}R_\nu I\delta\psi$$
$$\nabla^{-1}P_{k_4}[P_{k_5}\psi\nabla^{-1}(1-I)Q_{\nu j}[P_{k_6}\psi P_{k_7}\psi]]]||_{L_t^1\dot{H}^{-1}}$$
$$\leq C2^{-\delta_1|k_1-k_0|}2^{\delta_2[\min_{i=2,\dots,7}\{k_i\}-\max_{i=2,\dots,7}\{k_i\}]}\tilde{c}_{k_0}$$
$$||\nabla_{x,t}P_{k_0}Q_{\geq k_0}[P_{k_1}\delta\psi_1 P_{k_2}\nabla^{-1}[P_{k_3}R_\nu I\delta\psi$$
$$\nabla^{-1}P_{k_4}[P_{k_5}\psi\nabla^{-1}(1-I)Q_{\nu j}[P_{k_6}\psi P_{k_7}\psi]]]||_{\dot{X}_{k_0}^{-\frac{1}{2},-1,2}}$$
$$\leq C2^{-\delta_1|k_1-k_0|}2^{\delta_2[\min_{i=2,\dots,7}\{k_i\}-\max_{i=2,\dots,7}\{k_i\}]}\tilde{c}_{k_0}$$

The remaining error terms are treated analogously.

CHAPTER 4

# Proof of Theorem 2.3

We first check the algebra type estimate. Thus let $\psi_{1,2} \in \mathcal{S}(\mathbf{R}^{2+1})$; we need to estimate $||P_k[P_{k_1}\psi_1\nabla^{-1}P_{k_2}\psi_2]||_{S[k]}$. Of course we may assume that $k=0$. We decompose

$$P_0[P_{k_1}\psi_1\nabla^{-1}P_{k_2}\psi_2] = P_0Q_{<100}[P_{k_1}\psi_1\nabla^{-1}P_{k_2}\psi_2] + P_0Q_{\geq 100}[P_{k_1}\psi_1\nabla^{-1}P_{k_2}\psi_2]$$

We first consider the large modulation case, i. e. the 2nd summand on the right. Commence with the case $k_1 > 10$. We freeze the modulation of the output to dyadic size $\sim 2^l$, and further decompose into the following cases:

$$P_0Q_l[P_{k_1}\psi_1\nabla^{-1}P_{k_2}\psi_2] = P_0Q_l[P_{k_1}Q_{\geq l-10}\psi_1\nabla^{-1}P_{k_2}\psi_2]$$
$$+ P_0Q_l[P_{k_1}Q_{<l-10}\psi_1\nabla^{-1}P_{k_2}Q_{\geq l-10}\psi_2] + P_0Q_l[P_{k_1}Q_{<l-10}\psi_1\nabla^{-1}P_{k_2}Q_{<l-10}\psi_2]$$

We treat each of the summands on the right: for the first, observe that irrespective of whether $P_{k_1}\psi_1$ is of first or 2nd type,

$$||P_0Q_l[P_{k_1}Q_{\geq l-10}\psi_1\nabla^{-1}P_{k_2}\psi_2]||_{\dot{X}_0^{-(\frac{1}{2}-\mu),1-\mu,1}}$$
$$\leq C2^{(1-\mu)l}||P_{k_1}Q_{\geq l-10}\psi_1||_{L_t^2L_x^2}||\nabla^{-1}P_{k_2}\psi_2||_{L_t^\infty L_x^2}$$

When $l \geq k_1$, one estimates this by

$$\leq C\sum_{a\geq l-10} 2^{(1-\mu)(l-a)}||P_{k_1}Q_a\psi_1||_{\dot{X}_{k_1}^{-(\frac{1}{2}-\mu),(1-\mu),1}}||\nabla^{-1}P_{k_2}\psi_2||_{L_t^\infty L_x^2}$$

If, on the other hand, $l < k_1$, one estimates this by

$$\leq C2^{(\frac{1}{2}-\mu)l}||P_{k_1}\psi_1||_{\dot{X}_{k_1}^{0,\frac{1}{2},\infty}}||\nabla^{-1}P_{k_2}\psi_2||_{L_t^\infty L_x^2}$$

if $P_{k_1}\psi_1$ is of first type, and by

$$\leq C2^{(1-\mu)l}||P_{k_1}\psi_1||_{L_t^2L_x^{2+}}||P_{k_2}\nabla^{-1}\psi_2||_{L_t^\infty L_x^2}$$

if it is of 2nd type. Moreover, one estimates

$$||P_0[P_{k_1}\psi_1\nabla^{-1}\psi_2]||_{L_t^\infty L_x^{1+}} \leq C2^{-k_1}||P_{k_1}\psi_1||_{L_t^\infty L_x^2}||P_{k_2}\psi_2||_{L_t^\infty L_x^2}$$

Next, one estimates the output with respect to $||.||_{L_t^{1+}L_x^\infty}$ by interpolating between a crude estimate for $L_t^{1+}L_x^\infty$ gotten by placing the inputs into $L_t^{2+}L_x^\infty$ and a refined estimate for $||.||_{L_t^2L_x^\infty}$ by using improved $L_t^4L_x^\infty$-Strichartz norms for the inputs, resulting again in a small exponential gain in $k_1$. This yields the desired bound upon summing over $k_1 = k_2 + O(1) > O(1)$, showing that this contribution to the output is of 2nd type. Now consider the third summand in the above trichotomy:

we have $k_1 = k_2 + O(1) = l + O(1)$ in this case. First assume both $P_{k_{1,2}}\psi_{1,2}$ are of first type. We can decompose

$$P_0Q_l[P_{k_1}Q_{<l-10}\psi_1\nabla^{-1}P_{k_2}Q_{<l-10}\psi_2] = \sum_{\pm} P_0Q_l[P_{k_1}Q^{\pm}_{<l-10}\psi_1\nabla^{-1}P_{k_2}Q^{\pm}_{<l-10}\psi_2]$$

$$= \sum_{\pm}\sum_{\kappa_{1,2}\in K_{-k_1},\mathrm{dist}(\kappa_1,\kappa_2)\sim 1} P_0Q_l[P_{k_1,\kappa_1}Q^{\pm}_{<l-10}\psi_1\nabla^{-1}P_{k_2,\kappa_2}Q^{\pm}_{<l-10}\psi_2]$$

Now one estimates, using the definition of $S[k,\kappa]$:

$$||\sum_{\pm}\sum_{\kappa_{1,2}\in K_{-k_1},\mathrm{dist}(\kappa_1,\kappa_2)\sim 1} P_0Q_l[P_{k_1,\kappa_1}Q^{\pm}_{<l-10}\psi_1\nabla^{-1}P_{k_2,\kappa_2}Q^{\pm}_{<l-10}\psi_2]||_{L^2_tL^2_x}$$

$$\leq C\sum_{\pm}(\sum_{\kappa_1\in K_{-k_1}}||P_{k_1,\kappa_1}Q^{\pm}_{<l-10}\psi_1||^2_{PW[\pm\kappa_1]})^{\frac{1}{2}}(\sum_{\kappa_2\in K_{-k_1}}||P_{k_2,\kappa_2}Q^{\pm}_{<l-10}\nabla^{-1}\psi_2||^2_{NFA[\pm\kappa_1]})^{\frac{1}{2}}$$

This in turn can be bounded by

$$\leq Cl^2 2^{-l}||P_{k_1}\psi_1||_{A[k_1]}||P_{k_2}\psi_2||_{A[k_2]}$$

Multiplying by $2^{(1-\mu)l}$ results in an acceptable estimate. Now assume $P_{k_1}\psi_1$ is of 2nd type, say. In that case, we estimate

$$||P_0Q_l[P_{k_1}Q_{<l-10}\psi_1\nabla^{-1}P_{k_2}Q_{<l-10}\psi_2]||_{L^2_tL^2_x}$$
$$\leq C||P_{k_1}Q_{<l-10}\psi_1||_{L^2_tL^{2+}_x}||P_{k_2}\nabla^{-1}\psi_2||_{L^\infty_tL^2_x}$$

Multiplication with $2^{(1-\mu)l}$ again yields an acceptable bound. One estimates the output with respect to $||.||_{L^\infty_tL^1_x}$ as well as $||.||_{L^{1+}_tL^\infty_x}$ as before, showing that this contribution is of 2nd type as well. The 2nd term of the trichotomy is treated like the first. Now consider the case $k_1 \in [-10, 10]$. We decompose

$$P_0Q_{>100}\partial_t[P_{k_1}\psi_1\nabla^{-1}P_{k_2}\psi_2]$$
$$= P_0Q_{>100}[P_{k_1}\partial_t\psi_1\nabla^{-1}P_{k_2}\psi_2] + P_0Q_{>100}[P_{k_1}\psi_1R_0P_{k_2}\psi_2]$$

We claim that if $P_{k_1}\psi_1$ is of first type, so is the output. This is immediate when $P_{k_2}\psi_2$ is of 1st type. Now assume that $P_{k_2}\psi_2$ is of 2nd type. If $P_{k_1}\psi_1$ has modulation $> 2^{10}$, this is again immediate. In the opposite case, one calculates

$$||P_0Q_{>100}[P_{k_1}Q_{<10}\partial_t\psi_1\nabla^{-1}P_{k_2}\psi_2]||_{L^2_tL^2_x}$$
$$\leq C||P_{k_1}Q_{<10}\partial_t\psi_1||_{L^\infty_tL^2_x}||\nabla^{-1}P_{k_2}Q_{>10}\psi_2||_{L^2_tL^\infty_x}$$

Using the definition of $B[k_2]$, one checks that the summation over $k_2$ can be carried out. Next, we use the first bilinear property in theorem 2.3(which will be proved later independently of this) to calculate

$$||P_0Q_{>100}[P_{k_1}\psi_1R_0P_{k_2}\psi_2]||_{L^2_tL^2_x}$$
$$\leq C(\sum_{c\in C_{k_1,k_2-k_1}}||P_0Q_{>100}[P_cQ_{<10}\psi_1R_0Q_{>90}\psi_2]||^2_{L^2_tL^2_x})^{\frac{1}{2}}$$
$$\leq C2^{\delta(k_2-k_1)}||P_{k_1}\psi_1||_{A[k_1]}$$

One can sum over $k_2 < 15$, getting the desired bound. Control over $||.||_L$ again follows via the Sobolev embedding. If $P_{k_1}\psi_1$ is of 2nd type, so is the output. This is a simple repetition of arguments before. The case $k_1 < -10$ is more of the same, which finishes the large modulation case. Now consider $P_0Q_{<100}[P_{k_1}\psi_1\nabla^{-1}P_{k_2}\psi_2]$. First assume $k_1 > 10$, and both $P_{k_{1,2}}\psi_{1,2}$ of first type. In that case, we claim that the output will be of 2nd type. We need to check that it is controlled with respect to both $||.||_{L_t^\infty L_x^{1+}}$ as well as $||.||_{L_t^{1+} L_x^\infty}$. For the first norm, this is immediate from Bernstein's inequality. For the 2nd norm, one interpolates between a crude estimate for $L_t^{1+} L_x^\infty$ gotten by placing both $P_{k_{1,2}}\psi_{1,2}$ into $L_t^{2+} L_x^\infty$ and an estimate for $||.||_{L_t^2 L_x^\infty}$ gotten by using improved Strichartz type norms for the inputs. The remaining improved Strichartz type norms constituting $||.||_L$ are controlled from Bernstein's inequality. Now assume $P_{k_1}\psi_1$ is of 2nd type. Then the output will again be of 2nd type, as is easily verified by placing $P_{k_2}\psi_2$ into $L_t^\infty L_x^2$. This concludes the case $k_1 > 10$. Now assume $k_1 \in [-10, 10]$, in which case $k_2 < 15$. First assume $P_{k_1}\psi_1$ is of first type. We claim that then the output will be of first type, irrespective of the type of $P_{k_2}\psi_2$. Commence with the case when $P_{k_2}\psi_2$ is of first type. We need to check the various parts constituting $||.||_{A[0]}$. First, we consider $||.||_{\dot{X}_0^{0,\frac{1}{2},\infty}}$. Freeze the modulation of the output and decompose

$$P_0 Q_j [P_{k_1}\psi_1 \nabla^{-1} P_{k_2}\psi_2] = P_0 Q_j [P_{k_1} Q_{\geq j-10}\psi_1 \nabla^{-1} P_{k_2}\psi_2]$$
$$+ P_0 Q_j [P_{k_1} Q_{<j-10}\psi_1 \nabla^{-1} P_{k_2} Q_{\geq j-10}\psi_2] + P_0 Q_j [P_{k_1} Q_{<j-10}\psi_1 \nabla^{-1} P_{k_2} Q_{<j-10}\psi_2]$$

We start with the first term on the righthand side: estimate

$$||P_0 Q_j [P_{k_1} Q_{\geq j-10}\psi_1 \nabla^{-1} P_{k_2}\psi_2]||_{\dot{X}_0^{0,\frac{1}{2},\infty}}$$
$$\leq C 2^{\frac{j}{2}} ||P_{k_1} Q_{\geq j-10}\psi_1||_{L_t^2 L_x^2} ||\nabla^{-1} P_{k_2}\psi_2||_{L_t^\infty L_x^\infty},$$

and one easily bounds this by

$$||P_{k_1}\psi_1||_{A[k_1]} ||P_{k_2}\psi_2||_{A[k_2]}$$

Now consider the 2nd term in the above trichotomy. We can estimate this as follows:

$$||P_0 Q_j [P_{k_1} Q_{<j-10}\psi_1 \nabla^{-1} P_{k_2} Q_{\geq j-10}\psi_2]||_{\dot{X}_0^{0,\frac{1}{2},\infty}}$$
$$\leq C ||P_{k_1} Q_{<j-10}\psi_1||_{L_t^\infty L_x^2} ||\nabla^{-1} P_{k_2} Q_{\geq j-10}\psi_2||_{L_t^2 L_x^\infty}$$
$$\leq C 2^{\min\{(\frac{1}{2}-\mu)(k_2-j),0\}} 2^{\min\{\frac{j-k_2}{4},0\}} ||P_{k_1}\psi_1||_{A[k_1]} ||P_{k_2}\psi_2||_{A[k_2]}$$

We have invoked the improved Bernstein's inequality [**37**], [**23**]. Note that one may sum here over $k_2 < 15$ for fixed $j < 100$. Now assume that $P_{k_2}\psi_2$ is of 2nd type. Then we don't use the preceding trichotomy, but divide into the cases $k_2 \geq j - 100$ and $k_2 < j - 100$. In the first case, we estimate

$$||P_0 Q_j [P_{k_1}\psi_1 \nabla^{-1} P_{k_2}\psi_2]||_{\dot{X}_0^{0,\frac{1}{2},\infty}} \leq C ||P_{k_1}\psi_1||_{L_t^\infty L_x^2} ||\nabla^{-1} P_{k_2}\psi_2||_{L_t^2 L_x^\infty}$$
$$\leq C 2^{\frac{j-k_2}{2}} ||P_{k_1}\psi_1||_{A[k_1]} ||P_{k_2}\psi_2||_{B[k_2]}.$$

In the case $k_2 < j - 100$, we split

$$P_0 Q_j [P_{k_1}\psi_1 \nabla^{-1} P_{k_2}\psi_2] = P_0 Q_j [P_{k_1} Q_{\geq j-10}\psi_1 \nabla^{-1} P_{k_2}\psi_2]$$
$$+ P_0 Q_j [P_{k_1} Q_{<j-10}\psi_1 \nabla^{-1} P_{k_2} Q_{\geq j-10}\psi_2]$$

The first summand on the right is estimated just as before:

$$||P_0 Q_j [P_{k_1} Q_{\geq j-10}\psi_1 \nabla^{-1} P_{k_2}\psi_2]||_{\dot{X}_0^{0,\frac{1}{2},\infty}}$$
$$\leq C 2^{\frac{j}{2}} ||P_{k_1} Q_{\geq j-10}\psi_1||_{L_t^2 L_x^2} ||\nabla^{-1} P_{k_2}\psi_2||_{L_t^\infty L_x^\infty}$$

as well as

$$||\sum_{k_2<j-100} P_0Q_j[P_{k_1}Q_{\geq j-10}\psi_1\nabla^{-1}P_{k_2}\psi_2]||_{\dot{X}_0^{0,\frac{1}{2},\infty}}$$
$$\leq C2^{\frac{j}{2}}||P_{k_1}Q_{\geq j-10}\psi_1||_{L_t^2L_x^2}||\nabla^{-1}\psi_2||_{L_t^\infty L_x^\infty}$$

As for the 2nd summand, we have

$$||P_0Q_j[P_{k_1}Q_{<j-10}\psi_1\nabla^{-1}P_{k_2}Q_{\geq j-10}\psi_2]||_{\dot{X}_0^{0,\frac{1}{2},\infty}}$$
$$\leq C2^{\frac{j}{2}}||P_{k_1}Q_{<j-10}\psi_1||_{L_t^\infty L_x^2}2^{(\frac{1}{2}-\mu)k_2}2^{-j(1-\mu)}||\nabla^{-1}P_{k_2}Q_{\geq j-10}\psi_2||_{\dot{X}_{k_2}^{-(\frac{1}{2}-\mu),1-\mu,1}}$$

Thus we can estimate this contribution by

$$\leq C2^{(\frac{1}{2}-\mu)(k_2-j)}||P_{k_1}\psi_1||_{A[k_1]}||P_{k_2}\psi_2||_{B[k_2]}$$

Observe that we have obtained the gain $2^{\frac{\min\{k_2-j,0\}}{2}}2^{(\frac{1}{2}-\mu)(j-k_2)}$, which allows us to sum over $k_2<15$. This concludes the estimate for $||.||_{\dot{X}_0^{0,\frac{1}{2},\infty}}$. We still need to control the complicated null-frame part, as well as $||.||_L$. Thus fix $l<-10$, and $-10\geq\lambda\geq l$, and consider an expression

$$|\lambda|^{-1}(\sum_{\kappa\in K_l}\sum_{R\in C_{0,\kappa,\lambda}}||\tilde{P}_RQ^{\pm}_{<2l}[P_{k_1}\psi_1\nabla^{-1}P_{k_2}\psi_2]||^2_{S[0,\pm\kappa]})^{\frac{1}{2}}$$

The following estimates are irrespective of the type of $P_{k_2}\psi_2$. Then we split this into a bunch of contributions: Observe the identity

$$\tilde{P}_RQ^{\pm}_{<2l}[P_{k_1}\psi_1\nabla^{-1}P_{<2l}Q_{<2l}\psi_2]=\tilde{P}_RQ^{\pm}_{<2l}[\tilde{P}_{R_1}Q^{\pm}_{<2l+O(1)}\psi_1\nabla^{-1}P_{<2l}Q_{<2l}\psi_2],$$

where $R_1=(1+\frac{1}{1000})R$. Next, note that from the definition of $S[k,\kappa]$, we have

$$||\tilde{P}_{R_1}Q^{\pm}_{<2l+O(1)}\psi_1||_{S[k,\pm\kappa]}\leq C\sum_{\kappa_1\in K_{l-100},\kappa_1\subset(1+\frac{1}{100})\kappa}||\tilde{P}_{R_1}P_{k_1,\kappa_1}Q^{\pm}_{<2l+O(1)}\psi_1||_{S[k,\pm\kappa]}$$
$$\leq C\sum_{\kappa_1\in K_{l-100},\kappa_1\subset(1+\frac{1}{100})\kappa}||\tilde{P}_{R_1}P_{k_1,\kappa_1}Q^{\pm}_{<2l+O(1)}\psi_1||_{S[k,\pm\kappa_1]},$$

Note that $\frac{11}{10}\kappa_1\subset\frac{11}{10}\kappa=\tilde{\kappa}$, the latter as in the definition of $PW[\kappa]$. Hence the above inequality. Putting these observations together, we get

$$|\lambda|^{-1}(\sum_{\kappa\in K_l}\sum_{R\in C_{0,\kappa,\lambda}}||\tilde{P}_RQ^{\pm}_{<2l}[P_{k_1}\psi_1\nabla^{-1}P_{<2l}Q_{<2l}\psi_2]||^2_{S[0,\pm\kappa]})^{\frac{1}{2}}$$
$$\leq C|\lambda|^{-1}(\sum_{\kappa\in K_{l+O(1)}}\sum_{R\in C_{0,\kappa,\lambda+O(1)}}||\tilde{P}_RQ^{\pm}_{<2l}\psi_1||^2_{S[k,\pm\kappa]})^{\frac{1}{2}}||\nabla^{-1}\psi_2||_{L_t^\infty L_x^\infty},$$

which leads to the desired estimate. Further, using

$$||P_0Q_{<2l}[P_{k_1}\psi_1\nabla^{-1}P_{<2l}Q_{\geq 2l}\psi_2]||_{\dot{X}_0^{0,\frac{1}{2},1}}$$
$$+||P_0Q_{<2l}[P_{k_1}\psi_1\nabla^{-1}P_{\geq 2l}\psi_2]||_{\dot{X}_0^{0,\frac{1}{2},1}}\leq C||P_{k_1}\psi_1||_{A[k_1]}$$

We can estimate

$$|\lambda|^{-1}(\sum_{\kappa\in K_l}\sum_{R\in C_{0,\kappa,\lambda}}||\tilde{P}_RQ^{\pm}_{<2l}[P_{k_1}\psi_1\nabla^{-1}P_{\geq 2l}\psi_2]||^2_{S[0,\pm\kappa]})^{\frac{1}{2}}$$
$$\leq C||P_0Q^{\pm}_{<2l}[P_{k_1}\psi_1\nabla^{-1}P_{\geq 2l}\psi_2]||_{\dot{X}_0^{0,\frac{1}{2},1}},$$

$$|\lambda|^{-1}\Big(\sum_{\kappa\in K_l}\sum_{R\in C_{0,\kappa,\lambda}}||\tilde{P}_R Q^{\pm}_{<2l}[P_{k_1}\psi_1\nabla^{-1}P_{<2l}Q_{\geq 2l}\psi_2]||^2_{S[0,\pm\kappa]}\Big)^{\frac{1}{2}}$$
$$\leq C||P_0 Q^{\pm}_{<2l}[P_{k_1}\psi_1\nabla^{-1}P_{<2l}Q_{\geq 2l}\psi_2]||_{\dot{X}_0^{0,\frac{1}{2},1}}$$

and so the desired estimate follows easily for the contributions of $P_{k_1}\psi_1\nabla^{-1}P_{\geq 2l}\psi_2$, $P_{k_1}\psi_1\nabla^{-1}P_{<2l}Q_{\geq 2l}\psi_2$. The estimate for $||.||_L$ is quite similar. This concludes the estimates when $P_{k_1}\psi_1$ is of first type. If it is of 2nd type, so will be the output. This is straightforward to check. The case $k_1 < -10$ is a tedious reiteration of similar estimates and hence omitted. This concludes the proof for the assertions of theorem 2.3 as far as the estimates concerning $||.||_S$ are concerned. We now proceed to the assertions concerning the bilinear estimates, as well as the estimates for

$$||R_0[\psi_1 A(\nabla^{-1}\psi_2)]||_{L_t^\infty L_x^2},\ ||P_k Q_{<k+O(1)}[\psi_1 A(\nabla^{-1}\psi_2)]||_{\dot{X}_k^{0,\frac{1}{2},\infty}}$$

We start with the latter, which follows from the refined assertion for the function $\beta$ in the decomposition

$$P_k Q_{<k+O(1)}[\psi_1 A(\nabla^{-1}\psi_2)] = \alpha+\beta$$

To understand this decomposition, one expands $A(\nabla^{-1}\psi_2)$ into a Taylor series (using real analyticity). One winds up with schematic expressions of the form $\psi\nabla^{-1}\psi$, $\psi\nabla^{-1}\psi\nabla^{-1}\psi$ etc. For the first type of expression, $P_k[\psi\nabla^{-1}\psi]$, the previous proof revealed that the only contribution of 2nd type arises from high-high interactions. But for these it is straightforward to verify that they are controlled with respect to $||.||_{\dot{X}_k^{0,\frac{1}{2},1}}$, see 3.4(a). Now one proceeds inductively, assuming the assertion to be true for both $P_{k_{1,2}}\psi_{1,2}$ $\forall k_{1,2}\in\mathbf{Z}$, and considering $P_k[P_{k_1}\psi_1\nabla^{-1}P_{k_2}\psi_2]$. For example, considering high-high interactions, when $P_{k_1}\psi_1$ is of 2nd type, one estimates

$$||P_k Q_{<k+O(1)}[P_{k_1}\psi_1\nabla^{-1}P_{k_2}\psi_2]||_{\dot{X}_k^{0,\frac{1}{2},1}} \leq C2^{(\frac{3}{2}-\epsilon)k}||P_{k_1}\psi_1||_{L_t^2 L_x^{2+}}||\nabla^{-1}P_{k_2}\psi_2||_{L_t^\infty L_x^2}$$
$$\leq C2^{(\frac{3}{2}-\epsilon)(k-k_2)}||P_{k_1}\psi_1||_{A[k_1]}||P_{k_2}\psi_2||_{S[k_2]}$$

One can sum here over $k_1 = k_2+O(1)\geq k+O(1)$, obtaining the desired estimate. In case of high-low interactions, one reasons as follows: assume $P_{k_1}\psi_1$ is of 2nd type. Then for $j<k+O(1)$, $k=k_1+O(1)$, we have

$$P_kQ_j[P_{k_1}\psi_1\nabla^{-1}\psi_2] = P_kQ_j[P_{k_1}Q_{\geq j-10}\psi_1\nabla^{-1}P_{<j-10}Q_{<j-10}\psi_2]$$
$$+P_kQ_j[P_{k_1}\psi_1\nabla^{-1}P_{<j-10}Q_{\geq j-10}\psi_2]+P_kQ_j[P_{k_1}\psi_1\nabla^{-1}P_{\geq j-10}\psi_2]$$

We estimate each of the terms on the right: for the first, we have

$$||\sum_{j<k+O(1)}P_kQ_j[P_{k_1}Q_{\geq j-10}\psi_1\nabla^{-1}P_{<j-10}Q_{<j-10}\psi_2]||_{\dot{X}_k^{0,\frac{1}{2},1}}$$
$$\leq C\sum_{j<k+O(1)}2^{\frac{j}{2}}\sum_{a\geq j-10}||P_{k_1}Q_a\psi_1||_{L_t^2L_x^2}||\nabla^{-1}\psi_2||_{L_t^\infty L_x^\infty}$$
$$\leq C||P_{k_1}\psi_1||_{\dot{X}_{k_1}^{0,\frac{1}{2},1}}||\nabla^{-1}\psi_2||_{L_t^\infty L_x^\infty}$$

Now consider the 2nd term on the right. We exploit the improved range of Strichartz type norms available for $P_{k_1}\psi_1$:

$$||P_kQ_j[P_{k_1}\psi_1\nabla^{-1}P_{<j-10}Q_{\geq j-10}\psi_2]||_{\dot{X}_k^{0,\frac{1}{2},\infty}}$$
$$\leq C2^{\frac{j}{2}}||P_{k_1}\psi_1||_{L_t^2L_x^{2+}}||\nabla^{-1}P_{<j-10}Q_{\geq j-10}\psi_2||_{L_t^\infty L_x^M}$$
$$\leq C2^{\frac{j-k_1}{2+}}||P_{k_1}\psi_1||_{A[k_1]}$$

One can sum over $j < k+O(1)$, resulting in the desired bound. Finally, the last term in the above trichotomy is handled similarly. One handles low-high interactions analogously. This shows that the desired property for $\beta$ is inherited from one stage of the expansion to the next.
Next, consider $P_kR_0[\psi_1A(\nabla^{-1}\psi_2)]$. One expands $A(\nabla^{-1}\psi_2)$ into a Taylor series, and proceeds inductively. Assume one has $\sup_{k\in\mathbf{Z}}||P_kR_0\psi_{1,2}||_{L_t^\infty L_x^2} \leq C$, and consider $P_kR_0[\psi_1\nabla^{-1}\psi_2]$. One rewrites this as

$$P_kR_0[\psi_1\nabla^{-1}\psi_2] = \sum_{k_1>k+10,k_1=k_2+O(1)} P_kR_0[P_{k_1}\psi_1\nabla^{-1}P_{k_2}\psi_2]$$
$$+ P_kR_0[P_{[k-10,k+10]}\psi_1\nabla^{-1}\psi_2] + P_kR_0[P_{<k-10}\psi_1\nabla^{-1}P_{[k-10,k+10]}\psi_2]$$

One treats each of these terms separately: for the first, let the derivative $\partial_t$ fall inside, replacing this by

$$||\sum_{k_1>k+10,k_1=k_2+O(1)} P_k\nabla^{-1}[P_{k_1}\partial_t\psi_1\nabla^{-1}P_{k_2}\psi_2 + P_{k_1}\psi_1R_0\psi_2]||_{L_t^\infty L_x^2}$$
$$\leq C\sum_{k_1=k_2+O(1)} [||P_{k_1}R_0\psi_1||_{L_t^\infty L_x^2}||P_{k_2}\psi_2||_{L_t^\infty L_x^2} + ||P_{k_1}\psi_1||_{L_t^\infty L_x^2}||R_0P_{k_2}\psi_2||_{L_t^\infty L_x^2}] \leq C$$

Next one estimates

$$||P_kR_0[P_{[k-10,k+10]}\psi_1\nabla^{-1}\psi_2]||_{L_t^\infty L_x^2}$$
$$\leq ||P_k\nabla^{-1}[P_{[k-10,k+10]}\partial_t\psi_1\nabla^{-1}\psi_2 + P_{[k-10,k+10]}\psi_1R_0\psi_2]||_{L_t^\infty L_x^2}$$
$$\leq C||P_{[k-10,k+10]}R_0\psi_1||_{L_t^\infty L_x^2}||\nabla^{-1}\psi_2||_{L_t^\infty L_x^\infty}$$
$$+ ||P_{[k-10,k+10]}\nabla^{-1}\psi_1||_{L_t^\infty L_x^2}||R_0P_{<k+15}\psi_2||_{L_t^\infty L_x^\infty}$$

Again one checks that this can be bounded by $C$. The third term above is more of the same. The assertion follows from this. We proceed to the bilinear estimates. For this we expand the expression
$P_k[\psi_1A(\nabla^{-1}\psi_2)]$ as a Taylor series, obtaining terms of the schematic form $\psi$, $\psi\nabla^{-1}\psi$, $\psi\nabla^{-1}\psi\nabla^{-1}\psi$ etc. Consider a typical such term of the form

$$P_0[\psi_1\nabla^{-1}\psi_2\nabla^{-1}\psi_3\ldots\nabla^{-1}\psi_a]$$

This term will have a coefficient decaying like $(a!)^{-1}$. Freeze the modulation of the output to dyadic size $\sim 2^l$. If $\nabla^{-1}\psi_a$ has modulation $\geq 2^{l+10}$, we can estimate

$$||P_0Q_lR_0[\psi_1\nabla^{-1}\psi_2\nabla^{-1}\psi_3\ldots\nabla^{-1}Q_{>l+10}\psi_a]||_{L_t^2L_x^2}$$
$$\leq C2^l||Q_{>l+O(1)}[\psi_1\nabla^{-1}\psi_2\nabla^{-1}\psi_3\ldots\nabla^{-1}\psi_{a-1}]||_{L_t^2L_x^2}||\nabla^{-1}Q_{>l+10}\psi_a||_{L_t^2L_x^2+L_t^2L_x^\infty}$$
$$\leq C2^{-\frac{l}{2+}}$$

Summing over $l > O(1)$ results in the desired upper bound. Similarly, we have

$$||\sum_{l>O(1)} P_0 Q_l R_0[\psi_1 \nabla^{-1}\psi_2 \nabla^{-1}\psi_3 \dots \nabla^{-1} Q_{[l-10,l+10]}\psi_a]||_{L_t^2 L_x^2}$$
$$\leq C(\sum_{l>O(1)} ||Q_{[l-10,l+10]} R_0 \psi_a||^2_{L_t^2 L_x^2 + L_t^2 L_x^\infty})^{\frac{1}{2}} \leq CC$$

which is again acceptable. Thus assume now that $\nabla^{-1}\psi_a$ has modulation $< 2^{l-10}$. If $[\psi_1 \dots \nabla^{-1}\psi_{a-1}]$ has modulation $\geq 2^{l-10}$, repeat the same process with this expression instead of the longer one. Continuing in this fashion, one either eventually[1] forces $\psi_1$ to be at modulation $> 2^{l-10a}$, or else one arrives at a situation of the following sort:

$$P_0 Q_l [Q_{>l-10(a-m-1)}[Q_{<l-10(a-m)}[\psi_1 \dots \nabla^{-1}\psi_m]\nabla^{-1}\psi_{m+1}] \dots \nabla^{-1}\psi_a]$$

If $\nabla^{-1}\psi_{m+1}$ has modulation $> 2^{l-10(a-m)}$, one argues just as before. One loses exponentially in $m$, which is counteracted by the small coefficient eventually applied to the expression from the Taylor expansion. Thus we may also apply an operator $Q_{<l-10(a-m)}$ in front of $\nabla^{-1}\psi_{m+1}$. In this case, however, we can write

$$P_0 Q_l [Q_{>l-10(a-m-1)}[Q_{<l-10(a-m)}[\psi_1 \dots \nabla^{-1}\psi_m]\nabla^{-1} Q_{<l-10(a-m)}\psi_{m+1}] \dots \nabla^{-1}\psi_a]$$
$$= P_0 Q_l [Q_{>l-10(a-m-1)}[P_{l-10(a-m)+O(1)} Q_{<l-10(a-m)}[\psi_1 \dots \nabla^{-1}\psi_m]$$
$$\nabla^{-1} P_{l-10(a-m)+O(1)} Q_{<l-10(a-m)}\psi_{m+1}] \dots \nabla^{-1}\psi_a]$$

As to the first bilinear inequality in Theorem 2.3, we decompose

$$P_k\phi = P_k Q_{\geq k+100}\phi + \sum_{\pm} \sum_{\kappa \in K_{-100}} P_{k,\kappa} Q^{\pm}_{<k+100}\phi$$

Then we have

$$(\sum_{c \in C_{k,r}} ||P_0 Q_l R_0 [Q_{>l-10(a-m-1)}[P_{l-10(a-m)+O(1)} Q_{<l-10(a-m)}[\psi_1 \dots \nabla^{-1}\psi_m]$$
$$\nabla^{-1} P_{l-10(a-m)+O(1)} Q_{<l-10(a-m)}\psi_{m+1}] \dots \nabla^{-1}\psi_k] P_c Q_{\geq k+100}\phi||^2_{L_t^2 L_x^2})^{\frac{1}{2}}$$
$$\leq C 2^{\min\{k+r,0\}} ||P_0 Q_l R_0 [Q_{>l-10(a-m-1)}[P_{l-10(a-m)+O(1)} Q_{<l-10(a-m)}[\psi_1 \dots \nabla^{-1}\psi_m]$$
$$\nabla^{-1} P_{l-10(a-m)+O(1)} Q_{<l-10(a-m)}\psi_{m+1}] \dots \nabla^{-1}\psi_k]||_{L_t^\infty L_x^2} ||P_k Q_{\geq k+100}\phi||_{L_t^2 L_x^2}$$
$$\leq C 2^{\min\{\frac{k}{2},0\}} 2^r 2^{10m}$$

The loss in $m$ will be counteracted by the small Taylor coefficients. Now consider the contribution of

$$\sum_{\pm} \sum_{\kappa \in K_{-100}} P_{k,\kappa} Q^{\pm}_{<k+100}\phi$$

[1]We may assume $a << l$, since otherwise one gets a large gain in $l$ just from the Taylor coefficient.

We have the identity

$$P_0Q_l[Q_{>l-10(a-m-1)}[P_{l-10(a-m)+O(1)}Q_{<l-10(a-m)}[\psi_1\dots\nabla^{-1}\psi_m]$$
$$\nabla^{-1}P_{l-10(a-m)+O(1)}Q_{<l-10(a-m)}\psi_{m+1}]\dots\nabla^{-1}\psi_k]$$
$$=\sum_{\pm}\sum_{\kappa_{1,2}\in K_{-100},\,\mathrm{dist}(\kappa_1,\kappa_2)\sim 1}$$
$$P_0Q_l[Q_{>l-10(a-m-1)}[P_{l-10(a-m)+O(1),\kappa_1}Q^{\pm}_{<l-10(a-m)}[\psi_1\dots\nabla^{-1}\psi_m]$$
$$\nabla^{-1}P_{l-10(a-m)+O(1),\kappa_2}Q^{\pm}_{<l-10(a-m)}\psi_{m+1}]\dots\nabla^{-1}\psi_k]$$

We may thus assume that either $\pm\kappa_1$ or $\pm\kappa_2$ has angular separation $\sim 1$ from $\pm\kappa$. Assume w. l. o. g. that $\mathrm{dist}(\pm\kappa_1,\pm\kappa)\sim 1$. Then estimate

$$(\sum_{c\in C_{k,r}}||P_0Q_lR_0[Q_{>l-10(a-m-1)}[P_{l-10(a-m)+O(1)}Q_{<l-10(a-m)}[\psi_1\dots\nabla^{-1}\psi_m]$$
$$\nabla^{-1}P_{l-10(a-m)+O(1)}Q_{<l-10(a-m)}\psi_{m+1}]\dots\nabla^{-1}\psi_k]P_cP_{k,\kappa}Q^{\pm}_{<k+100}\phi||^2_{L^2_tL^2_x})^{\frac{1}{2}}$$
$$\leq C2^l||[P_{l-10(a-m)+O(1),\kappa_1}Q^{\pm}_{<l-10(a-m)}[\psi_1\dots\nabla^{-1}\psi_m]||_{NFA^*[\pm\kappa]}$$
$$||\nabla^{-1}P_{l-10(a-m)+O(1)}Q_{<l-10(a-m)}\psi_{m+1}||_{L^\infty_tL^2_x}(\sum_{c\in C_{k,r}}||P_cP_{k,\kappa}Q^{\pm}_{<k+100}\phi||^2_{PW[\pm\kappa]})^{\frac{1}{2}}$$
$$\leq C2^{\frac{k}{2}}2^{\frac{r}{2+}}2^l||\nabla^{-1}P_{l-10(a-m)+O(1)}Q_{<l-10(a-m)}\psi_{m+1}||_{L^\infty_tL^2_x}$$

We have used that

$$||[P_{l-10(a-m)+O(1),\kappa_1}Q^{\pm}_{<l-10(a-m)}[\psi_1\dots\nabla^{-1}\psi_m]||_{NFA^*[\pm\kappa]}$$
$$\leq C||[P_{l-10(a-m)+O(1),\kappa_1}Q^{\pm}_{<l-10(a-m)}[\psi_1\dots\nabla^{-1}\psi_m]||_{\dot{X}^{\frac{1}{2},1}_{l-10(a-m)}+A[l-10(a-m)]}\leq C$$

One can sum now over $l$, obtaining the desired estimate with a loss $2^{10m}$, which is made up for by the small Taylor coefficient in front. The 2nd bilinear inequality is proved similarly, as is the version when $A(\nabla^{-1}\psi_2)$ is replaced by $A(\nabla^{-1}(\psi_2\nabla^{-1}\psi_3))$. This completes the proof of theorem 2.3.

# Bibliography


[1] P. D'Ancona, V.Georgiev *On the continuity of the solution operator of the wave maps system*, preprint

[2] P. Bizon, Comm.Math.Phys.215(2000), 45

[3] Cazenave, Thierry; Shatah, Jalal; Tahvildar-Zadeh, A. Shadi *Harmonic Maps and the development of singularities in Wave Maps and Yang-Mills fields.*, Ann. Inst. H. Poincare Phys. Theor. 68 (1998), no.3, 315-349

[4] D.Christodoulou, A. Tahvildar-Zadeh *On the regularity of spherically symmetric wave maps*, C.P.A.M., 46(1993), 1041-1091

[5] D. Christodoulou, A. Tahvildar-Zadeh *On the asymptotic behavior of spherically symmetric wave maps*, Duke Math. J.71 (1993), no.1, 31-69

[6] M. Guenther, *Isometric embeddings of Riemannian manifolds*, Proceedings Interntl. Congress of Mathematicians, Vol. 1-2 (Kyoto 1990),p.1137-1143, Tokyo, 1991, Math. Soc. Japan.

[7] C.-H. Gu, *On the Cauchy problem for harmonic maps defined on two-dimensional Minkowski space*, Comm. Pure Appl. Math. 33: 727-737, 1980.

[8] F.Helein, *Regularite des applications faiblement harmoniques entre une surface et une varietee Riemanienne*, C.R.Acad.Sci.Paris Ser.1 Math 312(1991), 591-596

[9] Keel, M.;, Tao,T. *Endpoint Strichartz estimates*, Amer.J. Math.120 (1998), no.5, 955-980

[10] S.Klainerman, *UCLA lectures on nonlin. wave eqns.*, preprint (2001)

[11] S.Klainerman, D.Foschi, *Bilinear Space-Time Estimates for Homogeneous Wave Equations*, Ann. Scient. Ec. Norm. Sup., 4e serie, t.33(2000), 211-274

[12] S.Klainerman, M.Machedon, *Smoothing estimates for null forms and applications*, Duke Math.J., 81(1995), 99-133

[13] S.Klainerman, M.Machedon, *On the algebraic properties of the* $H^{\frac{n}{2},\frac{1}{2}}$ *spaces*, I.M.R.N. 15(1998), 765-774

[14] S.Klainerman, M.Machedon, *On the regularity properties of a model problem related to wave maps*, Duke Math.J., 87(1997), 553-589

[15] S.Klainerman, I.Rodnianski, *On the global regularity of wave maps in the critical Sobolev norm*, I.M.R.N. 13(2001), 655-677

[16] S.Klainerman, S.Selberg, *Remark on the optimal regularity for equations of wave maps type*, C.P.D.E., 22(1997), 901-918

[17] S.Klainerman, S.Selberg, *The spaces* $H^{s,\theta}$ *and applications to nonlinear wave equations.*, preprint

[18] S.Klainerman, S.Selberg, *Bilinear estimates and applications to nonlinear wave equations*, preprint

[19] S.Klainerman, D.Tataru, *On the optimal regularity for the Yang-Mills equations in* $\mathbf{R}^{4+1}$, Journal of the American Math. Soc., 12(1999), 93-116

[20] J.Krieger, *Global Regularity of Wave Maps in 2 and 3 spatial dimensions*, Ph. D. Thesis, Princeton University (2003)

[21] J. Krieger, *Global regularity of Wave Maps from* $\mathbf{R}^{3+1}$ *to surfaces*, CMP 238/1-2 (2003), 333-366

[22] J.Krieger, *Null-Form estimates and nonlinear waves*, Adv. Differemtial Equations 8(2003), no.10, 1193-1236

[23] J. Krieger, *Global regularity of Wave Maps from* $\mathbf{R}^{2+1}$ *to* $\mathbf{H}^2$, CMP 250(2004), 507-580

[24] A.Nahmod, A.Stefanov, K.Uhlenbeck, *On the well-posedness of the wave maps problem in high dimensions*, Comm. Anal. Geom. 11(2003) no.1, 49-83

[25] S.Selberg,*Multilinear space-time estimates and applications to local existence theory for nonlinear wave equations*, Ph.D. thesis, Princeton University, 1999

[26] J.Shatah, A. Tahvildar-Zadeh, *On the Cauchy Problem for Equivariant Wave Maps*, Comm. Pure Appl. Math. 47(1994), 719-754

[27] J.Shatah, M.Struwe *The Cauchy problem for wave maps*, I.M.R.N.11(2002), 555-571

[28] J.Shatah, M. Struwe *Geometric Wave Equations*, AMS Courant Lecture Notes 2

[29] Sideris, Thomas,*Global existence of harmonic maps in Minkowski space*, Comm. Pure Appl. Math. 42(1989), no.1, 1-13

[30] Sogge, Christopher D., *Lectures on nonlinear wave equations*, Monographs in Analysis II, Interntl. Press, Boston, MA, 1995

[31] E. Stein, *Harmonic Analysis: Real-Variable Methods, Orthogonality and Oscillatory Integrals*, Princeton University Press, Princeton, NJ, 1993

[32] Sterbenz, J. *Angular regularity and Strichartz estimates for the wave equation*, preprint

[33] M.Struwe, *Equivariant Wave Maps in 2 space dimensions*, preprint

[34] M.Struwe, *Radially Symmetric Wave Maps from 1+2 dimensional Minkowski space to the sphere*, Math.Z.242(2002)

[35] T.Tao, *Ill-posedness for one-dimensional Wave Maps at the critical regularity*, Am. Journal of Math.122 No.3(200), 451-463

[36] T.Tao, *Global regularity of wave maps I*, I.M.R.N. 6(2001), 299-328

[37] T.Tao, *Global regularity of wave maps II*, Comm.Math.Phys.224(2001), 443-544

[38] T.Tao, *Counterexamples to the n=3 endpoint Strichartz estimate for the wave equation*, preprint

[39] D.Tataru, *Local and global results for wave maps I*, Comm. PDE 23(1998), 1781-1793

[40] D.Tataru,*On global existence and scattering for the wave maps equation*, Amer. Journal. Math.123(2001), no.1, 37-77

[41] D. Tataru, *Rough solutions for the Wave Maps equation*, preprint

## Editorial Information

To be published in the *Memoirs*, a paper must be correct, new, nontrivial, and significant. Further, it must be well written and of interest to a substantial number of mathematicians. Piecemeal results, such as an inconclusive step toward an unproved major theorem or a minor variation on a known result, are in general not acceptable for publication. Papers appearing in *Memoirs* are generally at least 80 and not more than 200 published pages in length. Papers less than 80 or more than 200 published pages require the approval of the Managing Editor of the Transactions/Memoirs Editorial Board.

As of January 31, 2006, the backlog for this journal was approximately 14 volumes. This estimate is the result of dividing the number of manuscripts for this journal in the Providence office that have not yet gone to the printer on the above date by the average number of monographs per volume over the previous twelve months, reduced by the number of volumes published in four months (the time necessary for preparing a volume for the printer). (There are 6 volumes per year, each containing at least 4 numbers.)

A Consent to Publish and Copyright Agreement is required before a paper will be published in the *Memoirs*. After a paper is accepted for publication, the Providence office will send a Consent to Publish and Copyright Agreement to all authors of the paper. By submitting a paper to the *Memoirs*, authors certify that the results have not been submitted to nor are they under consideration for publication by another journal, conference proceedings, or similar publication.

### Information for Authors

*Memoirs* are printed from camera copy fully prepared by the author. This means that the finished book will look exactly like the copy submitted.

The paper must contain a *descriptive title* and an *abstract* that summarizes the article in language suitable for workers in the general field (algebra, analysis, etc.). The *descriptive title* should be short, but informative; useless or vague phrases such as "some remarks about" or "concerning" should be avoided. The *abstract* should be at least one complete sentence, and at most 300 words. Included with the footnotes to the paper should be the 2000 *Mathematics Subject Classification* representing the primary and secondary subjects of the article. The classifications are accessible from `www.ams.org/msc/`. The list of classifications is also available in print starting with the 1999 annual index of *Mathematical Reviews*. The Mathematics Subject Classification footnote may be followed by a list of *key words and phrases* describing the subject matter of the article and taken from it. Journal abbreviations used in bibliographies are listed in the latest *Mathematical Reviews* annual index. The series abbreviations are also accessible from `www.ams.org/publications/`. To help in preparing and verifying references, the AMS offers MR Lookup, a Reference Tool for Linking, at `www.ams.org/mrlookup/`. When the manuscript is submitted, authors should supply the editor with electronic addresses if available. These will be printed after the postal address at the end of the article.

**Electronically prepared manuscripts.** The AMS encourages electronically prepared manuscripts, with a strong preference for $\mathcal{A}_{\mathcal{M}}\mathcal{S}$-LaTeX. To this end, the Society has prepared $\mathcal{A}_{\mathcal{M}}\mathcal{S}$-LaTeX author packages for each AMS publication. Author packages include instructions for preparing electronic manuscripts, the *AMS Author Handbook*, samples, and a style file that generates the particular design specifications of that publication series. Though $\mathcal{A}_{\mathcal{M}}\mathcal{S}$-LaTeX is the highly preferred format of TeX, author packages are also available in $\mathcal{A}_{\mathcal{M}}\mathcal{S}$-TeX.

Authors may retrieve an author package from e-MATH starting from `www.ams.org/tex/` or via FTP to `ftp.ams.org` (login as `anonymous`, enter username as password, and type `cd pub/author-info`). The *AMS Author Handbook* and the *Instruction Manual* are available in PDF format following the author packages link from `www.ams.org/tex/`. The author package can also be obtained free of charge by sending email to tech-support@ams.org (Internet) or from the Publication Division, American Mathematical Society, 201 Charles St., Providence, RI 02904-2294, USA. When requesting an author package, please specify $\mathcal{A}_{\mathcal{M}}\mathcal{S}$-LaTeX or $\mathcal{A}_{\mathcal{M}}\mathcal{S}$-TeX and the publication in which your paper will appear. Please be sure to include your complete mailing address.

**Sending electronic files.** After acceptance, the source file(s) should be sent to the Providence office (this includes any TeX source file, any graphics files, and the DVI or PostScript file).

Before sending the source file, be sure you have proofread your paper carefully. The files you send must be the EXACT files used to generate the proof copy that was accepted for publication. For all publications, authors are required to send a printed copy of their paper, which exactly matches the copy approved for publication, along with any graphics that will appear in the paper.

TeX files may be submitted by email, FTP, or on diskette. The DVI file(s) and PostScript files should be submitted only by FTP or on diskette unless they are encoded properly to submit through email. (DVI files are binary and PostScript files tend to be very large.)

Electronically prepared manuscripts can be sent via email to pub-submit@ams.org (Internet). The subject line of the message should include the publication code to identify it as a Memoir. TeX source files, DVI files, and PostScript files can be transferred over the Internet by FTP to the Internet node e-math.ams.org (130.44.1.100).

**Electronic graphics.** Comprehensive instructions on preparing graphics are available at www.ams.org/jourhtml/graphics.html. A few of the major requirements are given here.

Submit files for graphics as EPS (Encapsulated PostScript) files. This includes graphics originated via a graphics application as well as scanned photographs or other computer-generated images. If this is not possible, TIFF files are acceptable as long as they can be opened in Adobe Photoshop or Illustrator. No matter what method was used to produce the graphic, it is necessary to provide a paper copy to the AMS.

Authors using graphics packages for the creation of electronic art should also avoid the use of any lines thinner than 0.5 points in width. Many graphics packages allow the user to specify a "hairline" for a very thin line. Hairlines often look acceptable when proofed on a typical laser printer. However, when produced on a high-resolution laser imagesetter, hairlines become nearly invisible and will be lost entirely in the final printing process.

Screens should be set to values between 15% and 85%. Screens which fall outside of this range are too light or too dark to print correctly. Variations of screens within a graphic should be no less than 10%.

**Inquiries.** Any inquiries concerning a paper that has been accepted for publication should be sent directly to the Electronic Prepress Department, American Mathematical Society, 201 Charles St., Providence, RI 02904, USA.

## Editors

This journal is designed particularly for long research papers, normally at least 80 pages in length, and groups of cognate papers in pure and applied mathematics. Papers intended for publication in the *Memoirs* should be addressed to one of the following editors. In principle the Memoirs welcomes electronic submissions, and some of the editors, those whose names appear below with an asterisk (*), have indicated that they prefer them. However, editors reserve the right to request hard copies after papers have been submitted electronically. Authors are advised to make preliminary email inquiries to editors about whether they are likely to be able to handle submissions in a particular electronic form.

***Algebra** to ALEXANDER KLESHCHEV, Department of Mathematics, University of Oregon, Eugene, OR 97403-1222; email: ams@noether.uoregon.edu

**Algebra and its application** to MINA TEICHER, Emmy Noether Research Institute for Mathematics, Bar-Ilan University, Ramat-Gan 52900, Israel; email: teicher@macs.biu.ac.il

**Algebraic geometry** to DAN ABRAMOVICH, Department of Mathematics, Brown University, Box 1917, Providence, RI 02912; email: amsedit@math.brown.edu

***Algebraic number theory** to V. KUMAR MURTY, Department of Mathematics, University of Toronto, 100 St. George Street, Toronto, ON M5S 1A1, Canada; email: murty@math.toronto.edu

***Algebraic topology** to ALEJANDRO ADEM, Department of Mathematics, University of British Columbia, Room 121, 1984 Mathematics Road, Vancouver, British Columbia, Canada V6T 1Z2; email: adem@math.ubc.ca

**Combinatorics** to JOHN R. STEMBRIDGE, Department of Mathematics, University of Michigan, Ann Arbor, Michigan 48109-1109; email: FRS@umich.edu

**Complex analysis and harmonic analysis** to ALEXANDER NAGEL, Department of Mathematics, University of Wisconsin, 480 Lincoln Drive, Madison, WI 53706-1313; email: nagel@math.wisc.edu

***Differential geometry and global analysis** to LISA C. JEFFREY, Department of Mathematics, University of Toronto, 100 St. George St., Toronto, ON Canada M5S 3G3; email: jeffrey@math.toronto.edu

**Dynamical systems and ergodic theory** to AMIE WILKINSON, Department of Mathematics, Northwestern University, 2033 Sheridan Road, Evanston, IL 60208-2730; email: wilkinso@math.northwestern.edu

***Functional analysis and operator algebras** to MARIUS DADARLAT, Department of Mathematics, Purdue University, 150 N. University St., West Lafayette, IN 47907-2067; email: mdd@math.purdue.edu

***Geometric analysis** to TOBIAS COLDING, Courant Institute, New York University, 251 Mercer St., New York, NY 10012; email: traneditor@cims.nyu.edu

***Geometric analysis** to MLADEN BESTVINA, Department of Mathematics, University of Utah, 155 South 1400 East, JWB 233, Salt Lake City, Utah 84112-0090; email: bestvina@math.utah.edu

**Harmonic analysis, representation theory, and Lie theory** to ROBERT J. STANTON, Department of Mathematics, The Ohio State University, 231 West 18th Avenue, Columbus, OH 43210-1174; email: stanton@math.ohio-state.edu

***Logic** to STEFFEN LEMPP, Department of Mathematics, University of Wisconsin, 480 Lincoln Drive, Madison, Wisconsin 53706-1388; email: lempp@math.wisc.edu

***Ordinary differential equations, and applied mathematics** to PETER W. BATES, Department of Mathematics, Michigan State University, East Lansing, MI 48824-1027; email: bates@math.msu.edu

***Partial differential equations** to GUSTAVO PONCE, Department of Mathematics, South Hall, Room 6607, University of California, Santa Barbara, CA 93106; email: ponce@math.ucsb.edu

***Probability and statistics** to KRZYSZTOF BURDZY, Department of Mathematics, University of Washington, Box 354350, Seattle, Washington 98195-4350; email: burdzy@math.washington.edu

***Real analysis and partial differential equations** to DANIEL TATARU, Department of Mathematics, University of California, Berkeley, Berkeley, CA 94720; email: tataru@math.berkeley.edu

**All other communications to the editors** should be addressed to the Managing Editor, ROBERT GURALNICK, Department of Mathematics, University of Southern California, Los Angeles, CA 90089-1113; email: guralnic@math.usc.edu.

## Titles in This Series

856 **Vladimir Bolotnikov and Harry Dym,** On boundary interpolation for matrix valued Schur functions, 2006

855 **Yevgenia Kashina, Yorck Sommerhäuser, and Yongchang Zhu,** On higher Frobenius-Schur indicators, 2006

854 **Noam Greenberg,** The role of true finiteness in the admissible recursively enumerable degrees, 2006

853 **Joachim Krieger,** Stability of spherically symmetric wave maps, 2006

852 **Viorel Barbu, Irena Lasiecka, and Roberto Triggiani,** Tangential boundary stabilization of Navier-Stokes equations, 2006

851 **Jie Wu,** On maps from loop suspensions to loop spaces and the shuffle relations on the Cohen groups, 2006

850 **Siegfried Echterhoff, S. Kaliszewski, John Quigg, and Iain Raeburn,** A categorical approach to imprimitivity theorems for $C^*$-dynamical systems, 2006

849 **Katsuhiko Kuribayashi, Mamoru Mimura, and Tetsu Nishimoto,** Twisted tensor products related to the cohomology of the classifying spaces of loop groups, 2006

848 **Bob Oliver,** Equivalences of classifying spaces completed at the prime two, 2006

847 **Eric T. Sawyer and Richard L. Wheeden,** Hölder continuity of weak solutions to subelliptic equations with rough coefficients, 2006

846 **Victor Beresnevich, Detta Dickinson, and Sanju Velani,** Measure theoretic laws for lim–sup sets, 2006

845 **Ehud Friedgut, Vojtech Rödl, Andrzej Ruciński, and Prasad V. Tetali,** A Sharp threshold for random graphs with a monochromatic triangle in every edge coloring, 2006

844 **Amadeu Delshams, Rafael de la Llave, and Tere M. Seara,** A geometric mechanism for diffusion in Hamiltonian systems overcoming the large gap problem: Heuristics and rigorous verification on a model, 2006

843 **Denis V. Osin,** Relatively hyperbolic groups: Intrinsic geometry, algebraic properties, and algorithmic problems, 2006

842 **David P. Blecher and Vrej Zarikian,** The calculus of one-sided $M$-ideals and multipliers in operator spaces, 2006

841 **Enrique Artal Bartolo, Pierrette Cassou-Noguès, Ignacio Luengo, and Alejandro Melle Hernández,** Quasi-ordinary power series and their zeta functions, 2005

840 **Sławomir Kołodziej,** The complex Monge-Ampère equation and pluripotential theory, 2005

839 **Mihai Ciucu,** A random tiling model for two dimensional electrostatics, 2005

838 **V. Jurdjevic,** Integrable Hamiltonian systems on complex Lie groups, 2005

837 **Joseph A. Ball and Victor Vinnikov,** Lax-Phillips scattering and conservative linear systems: A Cuntz-algebra multidimensional setting, 2005

836 **H. G. Dales and A. T.-M. Lau,** The second duals of Beurling algbras, 2005

835 **Kiyoshi Igusa,** Higher complex torsion and the framing principle, 2005

834 **Kenichi Ohshika,** Kleinian groups which are limits of geometrically finite groups, 2005

833 **Greg Hjorth and Alexander S. Kechris,** Rigidity theorems for actions of product groups and countable Borel equivalence relations, 2005

832 **Lee Klingler and Lawrence S. Levy,** Representation type of commutative Noetherian rings III: Global wildness and tameness, 2005

831 **K. R. Goodearl and F. Wehrung,** The complete dimension theory of partially ordered systems with equivalence and orthogonality, 2005

830 **Jason Fulman, Peter M. Neumann, and Cheryl E. Praeger,** A generating function approach to the enumeration of matrices in classical groups over finite fields, 2005

829 **S. G. Bobkov and B. Zegarlinski,** Entropy bounds and isoperimetry, 2005

TITLES IN THIS SERIES

828 **Joel Berman and Paweł M. Idziak,** Generative complexity in algebra, 2005

827 **Trevor A. Welsh,** Fermionic expressions for minimal model Virasoro characters, 2005

826 **Guy Métivier and Kevin Zumbrun,** Large viscous boundary layers for noncharacteristic nonlinear hyperbolic problems, 2005

825 **Yaozhong Hu,** Integral transformations and anticipative calculus for fractional Brownian motions, 2005

824 **Luen-Chau Li and Serge Parmentier,** On dynamical Poisson groupoids I, 2005

823 **Claus Mokler,** An analogue of a reductive algebraic monoid whose unit group is a Kac-Moody group, 2005

822 **Stefano Pigola, Marco Rigoli, and Alberto G. Setti,** Maximum principles on Riemannian manifolds and applications, 2005

821 **Nicole Bopp and Hubert Rubenthaler,** Local zeta functions attached to the minimal spherical series for a class of symmetric spaces, 2005

820 **Vadim A. Kaimanovich and Mikhail Lyubich,** Conformal and harmonic measures on laminations associated with rational maps, 2005

819 **F. Andreatta and E. Z. Goren,** Hilbert modular forms: Mod $p$ and $p$-adic aspects, 2005

818 **Tom De Medts,** An algebraic structure for Moufang quadrangles, 2005

817 **Javier Fernández de Bobadilla,** Moduli spaces of polynomials in two variables, 2005

816 **Francis Clarke,** Necessary conditions in dynamic optimization, 2005

815 **Martin Bendersky and Donald M. Davis,** $V_1$-periodic homotopy groups of $SO(n)$, 2004

814 **Johannes Huebschmann,** Kähler spaces, nilpotent orbits, and singular reduction, 2004

813 **Jeff Groah and Blake Temple,** Shock-wave solutions of the Einstein equations with perfect fluid sources: Existence and consistency by a locally inertial Glimm scheme, 2004

812 **Richard D. Canary and Darryl McCullough,** Homotopy equivalences of 3-manifolds and deformation theory of Kleinian groups, 2004

811 **Ottmar Loos and Erhard Neher,** Locally finite root systems, 2004

810 **W. N. Everitt and L. Markus,** Infinite dimensional complex symplectic spaces, 2004

809 **J. T. Cox, D. A. Dawson, and A. Greven,** Mutually catalytic super branching random walks: Large finite systems and renormalization analysis, 2004

808 **Hagen Meltzer,** Exceptional vector bundles, tilting sheaves and tilting complexes for weighted projective lines, 2004

807 **Carlos A. Cabrelli, Christopher Heil, and Ursula M. Molter,** Self-similarity and multiwavelets in higher dimensions, 2004

806 **Spiros A. Argyros and Andreas Tolias,** Methods in the theory of hereditarily indecomposable Banach spaces, 2004

805 **Philip L. Bowers and Kenneth Stephenson,** Uniformizing dessins and Belyĭ maps via circle packing, 2004

804 **A. Yu Ol'shanskii and M. V. Sapir,** The conjugacy problem and Higman embeddings, 2004

803 **Michael Field and Matthew Nicol,** Ergodic theory of equivariant diffeomorphisms: Markov partitions and stable ergodicity, 2004

802 **Martin W. Liebeck and Gary M. Seitz,** The maximal subgroups of positive dimension in exceptional algebraic groups, 2004

801 **Fabio Ancona and Andrea Marson,** Well-posedness for general $2 \times 2$ systems of conservation law, 2004

For a complete list of titles in this series, visit the
AMS Bookstore at **www.ams.org/bookstore/**.